C000282170

Thinking Like a Plant

Thinking Like a Plant

A Living Science for Life

CRAIG HOLDREGE

Lindisfarne Books / 2013

Lindisfarne Books
An imprint of SteinerBooks/Anthroposophic Press, Inc.
610 Main Street, Great Barrington, MA 01230
www.steinerbooks.org

Copyright © 2013 by Craig Holdrege

All rights reserved. No part of this book may be reproduced in any form
without the prior written permission of the publisher, except for brief
quotations embodied in critical articles and reviews.

Library of Congress Cataloging-in-Publication Data
Holdrege, Craig, 1953-
  Thinking like a plant : a living science for life / Craig Holdrege. – 1st ed.
    p. cm.
  Includes bibliographical references and index.
  ISBN 978-1-58420-143-4 (pbk.) – ISBN 978-1-58420-144-1 (e-book)
  1. Plants–Philosophy. I. Title.
  QK46.H64 2013
  580.1–dc23
                          2013008209

Printed in the United States of America

Contents

Introduction

A PLANT GROWING UP through a crack in a sidewalk is embedded in a world of relations. Its roots grow downward toward the center of the earth. They explore and transform the soil while drinking up water and a fine array of minerals. The shoot and leaves grow in the opposite direction—upward into the air and light. The leaves open themselves to these elements and take them in. The plant brings the stream of water and minerals from the soil together with the air and light, and in its unassuming way does something miraculous—it creates its own living substance.

Living out of this embeddedness in the environment, the plant grows and transforms according to its own inner pattern and yet adjusts itself at all times to what it takes in from the environment. This means that a small and compact dandelion growing up through the crack may be strikingly different from one growing effusively as a "weed" only a few feet away in a flower bed.

Even though, as a creature of place, the plant is not mobile in the sense that an animal is, it is dynamic, connected, resilient, and, in its ever-changing life, always in relation to the world into which it grows. Why can't we be like that?

Clearly, we can't put much hope in our physical ability to wander out into the yard, take root, and then grow by using the sun's light and taking in air, water, and a paltry amount of minerals. But we can put hope in our ability to become plantlike thinkers. How might the world look if we as human beings were able to think the way a plant grows? Imagine gaining such flexibility of thought that our ideas were no longer rigid, static, and object-like, but grew, transformed, and, when necessary, died away. And as with plant form, what if our actions grew out of a context-sensitive relation to the world we inhabit? Isn't that a revolution worth striving for?

In this book I want to show that if we are interested in addressing our present-day unsustainable relation to the planet in a deep way, our way of thinking about the world needs radical reorientation. And as far as I can tell, there is no better model for sustainable human thinking than the plant. The problem is, our minds have fallen in love with objects and things. We have perfected what I will call object thinking. Object thinking takes for granted that nature consists of physical things and entities that interact on the basis of impersonal physical laws. It's a perspective that gives the intelligent intellectual mind a way to grasp nature as a complex mechanism or system and the ability to control and manipulate nature to a remarkable degree. But in becoming a mechanism or a system, nature also becomes a kind of abstraction. We are, in our minds, dealing with "things out there," a world of externality in which we really don't participate. Unless we are high-level scientists we don't have intimate relations with atoms, molecules, genes, hormones, or neurotransmitters, and yet we take it for granted that these "things"—of which we have no experience and to which we have no relation—are at the foundations of our world! As a result, who today has a sense of being vitally embedded within the living fabric of the world? I believe, and will argue, that object thinking is at the basis of our unsustainable relation to the world because it alienates us from the very world we are trying to understand and interact with in healthy, sustainable ways.

Because it is so deep-seated and pervasive, object thinking is a worldview that both exploiters of nature and environmental activists can share. From the object-thinking perspective, sustainability is a programmatic goal to be reached by means of already existing human capacities, and education is a means of training people (in traditional ways) to address pressing environmental issues. While such efforts are especially important in the short term, in the long term, and fundamentally, they will fall short. Long-term sustainability will not be achieved, I maintain, through technological fixes or environmental laws, as important as these may be. It demands an evolving state of mind—it would be better to say an evolving flux of mind—in which we experience ourselves as conscious participants in the planetary process and become increasingly able to model our ways of thinking and acting after the dynamic and interconnected nature of life itself. This is a tall order, because it means transcending object thinking, which has held its glorious and unglorious grip on the human mind far too long. But as with all large tasks, you seek a place to begin and get to work.

I want to show that it is possible to move beyond object thinking and develop what I will call living thinking. Living thinking is a participatory way of knowing that transcends the dichotomies of man-nature, subject-object, or mind-matter, which are so ingrained in the Western mind and form the bedrock of object thinking. One of my main guides in developing a participatory, transformative, and living way of relating to the world has been the work of the scientist and poet J. W. von Goethe. In the late eighteenth and first third of the nineteenth centuries Goethe carried out studies of plants, animals, color, and more. In all these endeavors his aim was to gain an understanding that reveals the life and vitality of the things he was studying. And he realized along the way that "if we want to behold nature in a living way, we must follow her example and make ourselves as mobile and flexible as nature herself" (Goethe, 2002, p. 56). He was deeply concerned with finding ways of adapting our sensibilities and conceptions to the things themselves. As a result science is not just about studying nature but entails the transformation of human consciousness.

One of the salient features of the Goethean approach is its striving to stay close to what is being studied. It values concreteness over abstraction. Through theories, models, and other mental constructs we place a human-made thought structure between ourselves and the things we are dealing with. These create distance between us and the world and become the primary context through which we see and assess it. So, for example, when we are dealing with life and ecology in the frame of object thinking, we think in terms of mechanisms. We attend to and find in the world—and, importantly, count as most real—that which corresponds to this mechanistic way of seeing. It is a different matter to ask of something in a much more open-ended way: what do you have to teach me? This question can become the beginning of a dialogue in which we strive to learn what the phenomena have to tell us—a process that will no doubt involve giving up a frame of mind we may have grown fond of and working to develop new and fresh perspectives. In discussing the Goethean approach from different angles in the various chapters of this book I will both show what we can do to begin to tread this pathway and address the obstacles that confront us on it. It is a matter of practice—finding new ways of perceiving, imagining, and thinking.

At The Nature Institute, a nonprofit research and education center in upstate New York that I direct, we have been practicing this approach in adult education since 2002. An overriding aim in our courses—mostly weeklong intensives—is to help people begin the shift from object thinking to a participatory living thinking. Many of the examples I give in this book are based on work we have done in these courses, and I include comments from participants in the narrative. An important initial motivation for me to write this book came out of conversations with people who had attended courses and asked, "Have you written about this so that I can continue the work on my own?"

The way I have crafted the book reflects my goal to allow the reader to enter into the processes of experience and discovery that lie at the heart of living thinking. It includes many images and concrete descriptions meant to facilitate the vital interweaving of thought and perception that is central to this way of engaging in the world. I want this book to be a practical guide for living thinking. As such I hope it opens up a traversable pathway for readers seeking a new relation to nature and can provide perspectives for educators on how education can become a truly participatory and experience-based process.

In chapter 1, I look at the nature of object thinking. By becoming conscious of object thinking and characterizing it, we begin to see its limited scope and its powerful propensity to objectify, reify, and manipulate. We realize that object thinking is only one way of dealing with the world—and not a simple reflection of "the way things are." This realization frees us to an extent from the fetters of object thinking and allows us to recognize that participation is a fundamental quality of human embeddedness in the world. We can choose to take the object-thinking frame of mind, but we can also choose to move beyond it. We are responsible for our way of knowing and can ask the question: how can our participation become more aligned with the world we are embedded in?

Once we begin to free our minds from object thinking, we can discover how much there is—in unexpected ways—to learn from living organisms. In this book I have chosen to focus—in chapters 2 through 5—on flowering plants. I address the question: can plants become our teachers of living thinking? I have a number of reasons for the focus on flowering plants. It would be hard to overestimate the value of turning our attention toward plant life as a means of becoming more ecocentric in our attention, especially since plants stand at the center of fundamental ecological processes on earth, processes upon which animal and human life depend. More pragmatically, I have studied flowering plants for many years, and most of our Nature Institute courses have revolved around plant study—wildflowers, trees, and plant communities—so that in my descriptions I can draw on those experiences. Moreover, Goethe himself and subsequent Goethean biologists have carried out extensive botanical studies upon which I can build. Since plants are everywhere and, unlike animals, don't run away from us, they allow us to engage with them in a phenomenological dialogue. Anyone who studies the examples in this book can, subsequently, at home on the window sill, out in the yard, in a neighboring park, or in an "empty" lot, find plenty of plants just waiting to reveal their secrets to the open and attentive human mind. Perhaps most important in our day and age is that plants are everything other than discrete, inert objects. It is as if they were destined to teach us to move beyond object thinking.

Viewed superficially, the seed is an independent object. But it bears within it the potential to develop into a complex organism. Under appropriate conditions—which differ from species to species—the seed begins to germinate. Its solid casing becomes porous to the moisture of the environment, the tissue swells, and physiological transformations occur; what was solid (starch) becomes fluid (sugar sap), the root grows out and down into the soil, while the shoot and leaves grow upward—

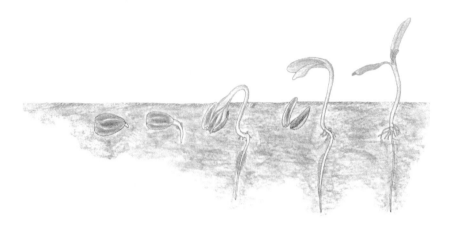

away from the center of the earth—and grow into the airy, light-filled environment. As we will see in chapter 2, in growth and development the plant is both receptive (open) and active. This is how it continually overcomes separation and generates life, a life that is stimulated and nourished by the world with which it connects. Or you could say: the stream of life in the plant can only be generated and maintained because the plant remains rooted in the world that sustains it. For us as human beings it is not a question of becoming physically rooted in the world. With our enclosed bodies we can move independently. We are more separate from the immediate life-giving environment that nourishes plants, and therefore also need plants as food to sustain us. But we can open ourselves and connect with the world as sentient perceiving and thinking beings. The question is: can we vitalize that connection so that we find our roots in the world through intensive engagement with what we perceive? In chapter 2, I will discuss a variety of practices that can help us develop our perceptual abilities as a means of participating in the life around us.

The plant has much to teach us about the nature of organic transformation, the topic of chapter 3. Its life history unfolds in a sequential and rhythmical way. While the roots are branching down into the soil, the leaves unfold along the stem. In many species of wildflowers there is a remarkable transformation in the form of the green leaves as they emerge one after another along the main stem: expansion of the small rounded leaves into the larger lobed and divided leaves, followed by a receding into small linear leaves. This contraction hints at more to come. Out of buds the flowers burst forth, revealing a new stage of development in the life of the plant. The flower forms a refined, ordered, and complex unit. In the center of the flower, the fruit develops and in it the seeds with their potential for abundant new plant life. Remarkably, the plant does not necessarily hold on to what it has already brought forth. Leaves often wilt and die

when the plant enters its flowering phase; the petals perish as the fruit develops, the fruit decays when the seeds are set free, and the seed case breaks open to allow the seedling to unfold. There is a rhythmic interplay of growth and decay in the life of the plant.

In following its development we come to know the plant as an organism that manifests itself over time. And we learn to see that in the dynamic process a unity is revealed. As Goethe expressed it, "The organ which expanded on the stem as leaf, assuming a variety of forms, is the same organ which now contracts in the calyx, expands again in the petal, contracts in the reproductive apparatus only to expand finally in the fruit" (1995, p. 96).

Getting to know the plant in this way lets us participate in essential qualities of life such as unfolding, growing and dying, transformation, dynamism, rhythm, and a unifying stream of creativity that brings forth diversity in an organism. Inasmuch as we internalize these qualities, our thinking itself can become enlivened. We can experience our own metamorphosis from an object perspective to a process perspective. We realize that each finished form is a snapshot in the life of an organism, and the plant can become a model for the way we work with our concepts and ideas. We can gain greater flexibility so that we do not hold on to our ideas in a rigid, static and object-like way. Rather, we can let them grow and transform, and, when appropriate, let the preliminary forms of our ideas die away. The plant can help us to establish a dynamic cognitive relation to the world. And when our thinking becomes more dynamic, it accesses a source of creativity—like the growing point in the plant—out of which new and fertile ideas can arise. As Lewis Mumford remarks, "Nothing will produce an effective change but the fresh transformation that has already begun in the human mind" (1970, p. 434). The plant can help us to effect that transformation.

I show through a variety of examples in chapter 4 how a plant's development and transformation is dependent on its relation to its environment. Plants develop their own form and substance through

the way they embed themselves in and respond to their environment. As the two different specimens of a mustard species shown here indicate, each plant forms itself in relation to the particular conditions it grows in. The one on the left grew in a composted garden soil, and developed a highly branched, dense root system, grew large, stout, and differentiated leaves, and formed many flowers. In contrast, when the same species grew in sandy loam soil (right), it stayed simpler—fewer roots, rounded and unlobed leaves, and few flowers. Each of the mustards not only shows us what species it is; it is also a disclosure of its environment. It is striking how each plant reveals in every part and as a whole the environment of which it is a part. Plants are not separate from their environment since we are seeing through them the qualities of the environment. Through revealing this greater unity of plant-environment, plants are showing us that life is eminently contextual, and they prod us, if we pay attention to them, to move beyond object thinking.

The ability of any given plant to form itself in connection with the conditions it meets in the world reveals its remarkable plasticity. It does not have a fixed program of development that it must follow under all circumstances. Rather, it has the ability to respond to a great array of conditions—and also to changing conditions that manifest during the course of its development—and then to enter into a kind of organic dialogue with those conditions. It is open to change and at the same time effects changes in the world in which it lives.

Through the way they live, plants provide a model for context-sensitive thinking. Instead of using the world as a probing ground for already-set agendas, instead of formulating hypotheses based on all-too-limited perspectives, instead of implementing programs to "fix" problems, we can gain the ability to enter into an open-ended, dynamic dialogue with the world in our thoughts and actions, so that increasingly they can reveal and enhance the living qualities of the world we inhabit. I conclude chapter 4 by discussing context sensitivity in scientific inquiry.

In chapter 5, I tell the story of a single plant, the common milkweed. In this portrayal we will meet a remarkable organism. I sketch its morphology, life history, and seasonal expressions. We meet an organism that is robust and vital, and yet also delicate and intricate. It forms manifold and unique relations to the insect world. It is so tightly entwined with insects that they could not live without the milkweed, just as the milkweed could not live without them. We learn in a concrete way that an organism extends beyond its apparent boundaries and is part of a greater, dynamic whole that we call its environment. And we come to know an organism as a specific and vital focus of relations in the world, unique and irreplaceable. This concrete engagement with the life of another organism, which entails coming into conversation with it from a variety of perspectives, kindles a deep awe and respect  for the wisdom of life on earth. And it gives real substance and gravity to the concept of biodiversity—to the thought that there are thousands upon thousands of such unique organisms, all intertwined in the dynamic and complex ecology of the earth.

Living thinking can vitalize every area of human engagement. And where would this be more important than in education? In the concluding chapter I highlight some of the perspectives, insights, and practices that emerge when, in the disposition of living thinking, we consider education.

One could hardly imagine a better teacher of dynamism, connectedness, resilience, and wholeness than the plant. The plant shows us how to live in transformation; it shows us context sensitivity; it shows us the unique nature of organisms; it shows us how to overcome an object-relation to the world. By developing a thinking modeled after such characteristics, our thinking becomes fluid and dynamic; we realize how we are embedded in the world; we become sensitive and responsive to the contexts we meet in the world; and we learn to thrive within a changing world. These are, I suggest, precisely the qualities our culture needs to develop a sustainable, life-supporting relation to the rest of the world.

It is easy to talk about all that is needed and to critique our current state of affairs. It is not easy to move beyond the status quo. I hope this book will be a practical and useful guide to developing a life-infused way of looking, thinking, and acting.

CHAPTER 1

From Object Thinking
to Living Thinking

IF WE DID NOT PERCEIVE the profound disconnect between the workings of nature and human thinking and action, we wouldn't be asking how we can inhabit the planet in a more sustainable way—not only for us but for the whole earth and its inhabitants. Again and again we see how our ideas that flow into actions create ecological problems. And today we see such effects at a global scale. We can hardly underestimate the degree to which we ignorantly and arrogantly impress our limited human perspective upon the rest of the earth.

And yet, we are part of that earth. We are not creatures from another planet. And so we are confronted with the odd predicament that we are beings connected with and sustained by the rest of the world and yet our actions lack connectedness with the ecological workings of this world. In many ways this situation seems intractable. This would only be the case, however, if we could not change. But we can.

The Objective Attitude

The solid ground beneath me gives me firm footing. Dwelling on the earth, I perceive myself as a body among other bodies, each with its own boundaries. When I run up against something, I know that what I am hitting is "not me." The rock I hold in my hand or the flower pot I put on the window sill each has its own existence, independent and separate from me. I have the experience of "me here" and "things out there." I take as a given that there is an objective world out there that is in principle separate from me.

It is also "self-evident" that the desk I am sitting at today is the desk I sat at yesterday. I perceive the world as largely stable even though at my desk, in the light conditions I find today, no detail looks precisely the same as yesterday. But I don't attend to those differences, or if I do, I deem them unessential. What I feel to be real and essential is the continuity from yesterday to today. My ability to remember and revive the past, while moving into the future, plays a significant role in this stable world.

Without this objective attitude we would be immersed and engaged in all our perceptions and doings, but would never be able to stand back and consider: "What did I just do?" Or: "Did that snake really have yellow streaks on its sides?" We can detach ourselves from immediate experience and reflect on it, as if from a distance. This is a gift we all rely on to make our way in the world. In many respects we'd be lost without it. It not only provides a kind of coherence for our day-to-day lives, but also helps us to navigate through uncertainties and problems we meet.

Most of us would feel decidedly uncomfortable in a wholly fluid world in which our day-to-day experience was of immersion in ongoing Heraclitean change, in which every experience was like a river into which we could never step twice. We are comfortable in a "state of mind in which one knows phenomena precisely in the act of distancing oneself from them," as Morris Berman puts it well in a study of modern consciousness (1984, p. 27). This pulling back is a form of disengagement and is a first important act of creating separation between ourselves and the rest of the world.

With this separation the world becomes a riddle for human consciousness. We ask questions, we ponder and think about our relation to things. These activities are both symptoms of our disconnection and at the same time ways of trying to bridge it. While there are many imaginable ways we could try to bridge this gap, in Western culture a particular style of thought has come to dominate the way we try to understand the riddles of the world. It is what I will call object thinking.[1]

Object Thinking

Water is something we all know well: For a child it's the wonderful substance that makes crumbly dirt into a beautiful thick soup to serve imaginary guests. Water lightens your weight, and its warmth soothes you in an evening bath. Water drops from the sky, splashes into the stream, and becomes part of the river system that connects glaciers with oceans. Water tumbles over a ledge, becomes the spray of millions of water droplets, only to so easily merge with the flowing body of water below. Water meanders through the landscape, as if avoiding straight lines at all costs. Water quenches thirst and enlivens. Water becomes solid ice when the cold grips it, only to reemerge as a fluid when enough warmth penetrates it again. And water miraculously allows water striders to dance on its surface (and you wish you could do the same).

With such experiences, and many others that have become part of who we are, we meet water in school. I don't mean spraying your classmates from the drinking fountain. I mean when things get serious in the classroom. If you're lucky, the teacher will take you outside to a creek and let you observe water striders or water beetles cruising around the surface of the water; she'll let you put different sizes and shapes of objects on the water and let you see what happens. Back in the classroom she'll let you float some paper clips on the surface of water and let you and your classmates discover if there are changes if you add some soap. If you are less lucky, she will show videos of such observations and experiments. And if you are even less lucky, she'll simply say, "You all know that water striders can skip around on the surface of water; how do they do that?"

Whether you have the fortune of first immersing yourself in an array of experiences or are simply confronted with a plump question,

Figure 1.1. Mating Water Striders
(Photo Markus Gayda)

at some point the teacher will get to what she believes—or what the curriculum guidelines say—is the meat of the matter. I mean the scientific explanation. How does water hold these creatures up? The creative teacher who wants to stimulate students to inquire themselves won't simply give an answer but will let them come up with their own suggestions. After further explorations she may introduce a scientific term, "surface tension." But usually there is another step: how do we *explain* surface tension? It is, somehow, not enough to explore the phenomena in a variety of manifestations: we believe we need an explanation.

The explanation goes something like this: Water consists of water molecules; we can picture them, for the sake of explanation, as little balls; these little balls are attracted to each other and form a kind of skin at the water's surface. Holding together, the molecules don't let the water strider's tiny feet penetrate. But when you put soap in the water the attraction between the molecules is lessened; therefore the surface tension is lessened, and the unlucky water strider that lands on dishwater sinks.

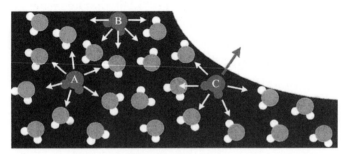

Figure 1.2. Schematic diagram of water molecules to illustrate surface tension. Forces are represented by arrows. Water molecules within the body of the water (A), at the surface (B), and at the surface that is being indented (for example, by a water strider).

What happens to water in this explanation? It becomes a composite of many little solids! The tiny interacting particles become the real constituents of water and somehow (don't ask how) all together they give the appearance of being the unified substance we know as "water." This is the moment of thinking I want to pay attention to. It is the shift from experiencing water and getting to know it in a variety of situations to explaining. And what an explanation entails has been determined beforehand: it is about coming up with a model that assumes physical entities and forces lie behind the phenomenon in question, and these are viewed as the cause of its appearance.

When we explain surface tension through the interaction of particles we are entering with our minds into the water. But we are not entering the water in the way of a water droplet that falls into the pond and merges with it. No, we enter the water mentally maintaining the form of the solid. We picture water as consisting of discrete little interacting entities, and in this way we can wrap our minds around water. Clearly, in so doing, we lose a good deal of what water is all about, but we also have a handle on it. It doesn't slip through our fingers. When we mentally analyze it into distinct entities, we can interpret characteristics in terms of those things. And things make sense to us. So once the unity of water becomes particles for our minds, we can construct models of their interaction based on mechanical and electrical forces. A logical and consistent model is the result, and it has the character of being transparent for my own thinking. As long as we don't feel the model drastically contradicts the characteristics of water we're focusing on, and even better, if it allows us to find and investigate other characteristics, the model solidifies in our minds and becomes for us increasingly the reality of water.

This way of explaining and trying to make sense of the world is deep-seated in our culture. It is exercised and taken for granted to such a degree that we hardly notice how strange it is to conceive of a fluid as a solid in order to make sense of it. And in the process we hardly notice how much of fluidity we have lost sight of.

For the moment it's not so important whether a given model stands the test of time or is soon discarded. It is the gesture of thought I'm interested in. In the objective attitude we create a distance between ourselves and the world and can consider it in a cool and calm way. In object thinking we take further steps away from the rich and often messy nature of the world. We conceive everything in object-like terms and, especially in science, project a more ordered, object-like world behind it. Such a model is transparent to thought and allows us to make sense of aspects of the world.

As French philosopher Henri Bergson describes, in object think-ing—he speaks of the intellect—we have "the natural obstinacy with which we treat the living like the lifeless and think all reality, however fluid, under the form of the sharply defined solid" (1998, p. 165). For this reason it seems "natural" for us today to entertain the notion that water consists of object-like molecules of hydrogen and oxygen, or that genes in DNA are the essential building blocks of life, just as atoms are the building blocks of molecules. It may be that only a select few thinkers and scientists come up with such notions and refine them to a remark-able degree, but once they are there the rest of us can accept them, and although we may not understand them, they don't raise our hackles.

Inasmuch as object thinking abstracts from the fullness of human experience and focuses on what it views as the underlying causes of phenomena, a remarkable—and remarkably one-sided and strange—picture of reality emerges. Here is a description of reality by the con-temporary philosopher Paul Churchland in his book *Matter and Consciousness:*

> The red surface of an apple does not look like a matrix of mole-cules reflecting photons at certain critical wave lengths, but that is what it is. The sound of a flute does not sound like a compression wave train in the atmosphere, but that is what it is. The warmth of the summer air does not feel like the mean kinetic energy of mil-lions of tiny molecules, but that is what it is. (1988, p. 15)

For Churchland, reality consists of the high-level, object-like abstrac-tions of science. The apples we see and taste, the melody we hear, and the warmth we sense are all only appearances, mere subjective sem-blances of true physical reality. And what about our own inwardness? Neuroscientist Antonio Damasio, writing in *Nature*, has an answer:

> An emotion, be it happiness or sadness, embarrassment or pride, is a patterned collection of chemical neural responses that is pro-duced by the brain when it detects the presence of an emotionally competent stimulus. (2001, p. 781)

The resulting worldview is one in which the world we actually experience becomes a numinous and basically unreal aftereffect of a "real world" that consists of molecules, photons, chemical neural responses, and the like. This "real world" is not one we directly experience but is a world that resides in our thought models, which have been constructed in concert with machine-mediated experimental findings.

You may find this perspective crass. And it is. You could say it is object thinking in its hardest form. In the biological sciences—the sciences of life—object thinking takes the form of the search for underlying mechanisms. "Mechanism" or "mechanistic explanation" are key words in most journal articles reporting on biological research. Every biology student learns early on, when confronted with an interesting and question-raising phenomenon, to always ask the question, what is the underlying mechanism? The question "What is life" becomes "What is a minimal set of essential cellular genes?" (Hutchinson et al., 1999).

The modern mind gazes, so to speak, through the phenomena to an assumed "real" world of interacting entities. Rudolf Steiner suggests that object thinking is imbued with a kind of thought inertia, since it tries "to pierce the veil of the senses and to construct something more behind it with the aid of our concepts.... I take my lesson from inert matter, which continues to roll on even when the propulsive forces have ceased" (1983, p. 22). In this sense, object thinking imitates the behavior of physical objects in the way it relates to the sense world. However, on the real earth, objects come to rest, while there is nothing to prohibit thoughts from simply rolling along.

When we take objects as the model entities of the world and view all phenomena as if they were things or consist of thing-like parts, we are using object thinking. One way of characterizing much of modern science is to say that it has perfected object thinking and applied it full force—excluding other ways of viewing things—to the investigation and manipulation of the world. Science is imbued with the virtues of the objective attitude—the ability to gain distance to experience and to consider things calmly, clearly, and in logical terms. But science takes another critical step when it strives to reduce all happenings in nature to merely material occurrences. Since our culture is science- and technology-dominated, object thinking has become an increasingly pervasive element in the modern psyche.

If I phrase this mode of thought as a verb, I can say that we continually and with insistence "thing" the world. When we "thing" the world, we view it as consisting of entities that are self-enclosed—having physical, spatial boundaries—and interact with one another in external ways.

The Logic of Solids

Object thinking not only informs how we view the things "out there," but also how we think about and form our concepts, a fact that has been noticed by a variety of thinkers. Bergson describes how "our concepts have been formed on the model of solids" and how "our logic is, pre-eminently, the logic of solids..." (1998, p. ix). For the intellectual mind,

> concepts, in fact, are outside each other, like objects in space; and they have the same stability as such objects, on which they have been modeled. Taken together, they constitute an "intelligible world," that resembles the world of solids in its essential characters, but whose elements are lighter, more diaphanous, easier for the intellect to deal with than the image of concrete things; they are not, indeed, the perception itself of things, but the representation of the act by which the intellect is fixed on them. They are, therefore, not images, but symbols. Our logic is the complete set of rules that must be followed in using symbols. (Bergson, 1998, pp. 160–161)

Or, in David Bohm's words: "Logically definable concepts play the same fundamental role in abstract and precise thinking as do separable objects and phenomena in our customary description of the world" (1989, p. 170). Think of the basic principles of logic:

- The principle of identity: if a statement is true, then it is true (A is A).
- The principle of noncontradiction: no statement can be both true and not true at the same time.
- The principle of excluded middle: a statement is either true or false.

All these principles follow the logic of solids—either/or relations and clear-cut boundaries. When Descartes, as one of the founders of both modern philosophy and mathematical physics, appealed to "clear and distinct ideas" as his measure of truthfulness (1968, Meditation III, p. 35), he was basing his philosophy on the logic of solids. A definition—clearly circumscribed, distinct, and transparent—is the epitome of object thinking. If you can define something, you "have" it.

So object thinking divides any phenomenon or process into discrete elements—it isolates a feature of reality, considers it as a thing, isolates something else as a thing, and then investigates how the things interact. Through the process of analysis and isolation, we gain precision and the ability to formulate quantitative relations. We also gain a power to predict, and, in what is judged to be the ultimate test and validation of object thinking, the ability to control. As Bergson writes, "The human intellect feels at home among inanimate objects, more especially among solids, where our action finds its fulcrum and our industry its tools" (1998, p. ix). So object thinking is both theoretically powerful and, through technology, exercises immense practical power.

Object Thinking Applied

Object thinking is not simply one way of viewing the world today. It is the canonical view, embodied in education, science, technology, and even in religion. While perfected in science, it is part of all of us. It reveals itself in the way we talk about how genes, molecules, hormones, or brain function "cause" form, function, and behavior. Object thinking flows into action and changes the world according to its perspective. The consequences of object thinking are immense since its propensity is to change the way people see themselves and the world and to largely determine the way modern societies interact with the rest of nature.

Since the birth of modern science, the object view of nature has stimulated research. With the mind attuned to the mechanical, not only did scientists begin to see the mechanical in nature everywhere, but mechanical devices aided the investigation of nature as mechanism (think of clocks, telescopes, and microscopes). Through experimentation you manipulate nature to answer your questions. Experimentation is itself a kind of technology, and science becomes ever more a technological endeavor. If you view nature as an object then, as far as manipulation and technology go, nature becomes a resource to be used. It has no "inside" that needs to be taken into account. As Carolyn Merchant points out in *The Death of Nature*:

> The removal of animistic, organic assumptions about the cosmos constituted the death of nature—the most far-reaching effect of the Scientific Revolution. Because nature was now viewed as a system of dead, inert particles moved by external, rather than inherent forces, the mechanical framework itself could legitimate the manipulation of nature. Moreover, as a conceptual framework, the mechanical world order had associated with it a framework of values based on power, fully compatible with the directions taken by commercial capitalism. (1983, p. 193)

Merchant says that capitalism is "fully compatible" with the object view of nature. It is not only compatible with it, it fully embraces and embodies it. As the atom or gene is the focus of mechanistic nature so is the individual capitalist the fundamental unit of capitalism around which everything revolves. Community is an epiphenomenon in a capitalist world, just as the organism is an epiphenomenon in a gene-driven world. Capitalism thinks in terms of market "mechanisms" that determine human behavior, with competition being a ruling force. And in capitalism everything becomes a commodity—an "object" that can be possessed and sold, including land, organisms, and labor.

There is, however, one dominant and powerful organic metaphor in capitalism: the idea of economic "growth." But growth here—and this is telling—means accumulation of wealth and an increase in the production of commodities. Key to the capitalist, free-market economy is the idea that such growth is boundless and that continuing growth is the key to a "healthy" economy. Although the word *growth* is drawn from life, in capitalist economics it is framed in terms of object thinking. Nowhere in healthy life do you find unlimited, linear growth. Rather you find bounded growth embedded within a context, and you find regulated processes of death and decay that accompany any growth process. The idea of a growth economy in capitalism is not an idea grounded in any understanding of healthy life processes. If you search in the organic world for something close to a correlate of economic growth, you find at best unchecked cancerous growth—in other words, a pathological phenomenon. That we let the economy inflate into a "prosperous" bubble that at some point collapses, after which we eagerly await (and expect) the saving grace of the next bubble (dot-com, housing) is an all-too-clear sign of the depth of pathology we are dealing with (Krugman, 2008).

When object thinking gets to work, it tries to solve problems: It tends to isolate an area of phenomena ("the problem"), to analyze it in detail, and to come up with a targeted, technological solution to the problem. But because it isolates, it does not take into account the larger, more dynamic context of any "problem," and the proposed solution will often cause its own set of problems. Amory Lovins, an expert and innovator in energy technologies, speaks out of a wealth of experience when he writes, "If you don't know how things are connected then the cause of problems is solutions" (2001). This is the problem of technological solutions "biting back," which has been well documented (see Tenner, 1997).

Let me bring one example of such a technological solution to a problem from an area I know well, genetic engineering (Holdrege & Talbott, 2008, chapter 2). Many children in the Third World go blind due to Vitamin A deficiency. The human body transforms beta-carotene into Vitamin A, which our eyes need for healthy vision. Since for most children in Asia white rice is a staple, biotechnologists came up with the idea of genetically modifying rice to accumulate beta-carotene in the rice kernel. In this way children would receive beta-carotene in their diets to prevent blindness. Biotechnologists have in fact succeeded in this manipulation of rice (Paine, et al., 2005; Ye, et al., 2000). This genetic alteration is presently being bred into typical Asian rice varieties. It is not yet on the market. Because its kernels are yellowish and because of its promise, this genetically modified rice has been given the name "golden rice."

Considered in isolation (object thinking), golden rice seems to make sense and to be a worthy program. But let's look at the larger context in a concrete way. I will touch on only a couple of aspects. First, beta-carotene does not get taken up by the gut unless there are adequate amounts of oil or fats in the diet. Its further transformation in the body and the functioning of Vitamin A in the retina are dependent on enzymes, which are in turn dependent on adequate amounts of proteins in the diet (Erdman, Beirer, & Gugger, 1993). So if the malnourished children eat golden rice, but do not receive at the same time sufficient oils or fats and protein in their diet, the extra beta-carotene will do them little good. We can't think of beta-carotene in isolation if we want our thinking to be in touch with reality. What's important is a *balanced* diet, not the addition of one factor. Second, "nearly eighty percent of all malnourished children in the developing world in the early 1990s lived in countries that boasted food surpluses" (Gardner & Halweil, 2000, p. 17). The problem is, therefore, not so much that there is not enough food, but that it is not available to the hungry. It's not getting to the people who need it most—the poor people who don't have enough money to buy enough food. So how is golden rice going to get to the targeted children, and if they get it, will it help them? Third, even if people are offered golden rice, the question is, would they eat it? Most people in Asia eat white rice instead of the healthier brown rice. This has to do with appearance, consistency, storage quality, and many other things. It is naïve to think that people will simply discard thousands of years of tradition for a new product. As Jacques Ellul writes, "We must not think that people who are the victims of famine will eat anything. Western people might, since they no longer have any beliefs or traditions or sense of the sacred. But not others. We have thus to destroy the whole social structure, for food is one of the structures of society" (Ellul 1990, p. 53).

This well-meaning "solution" to a problem suffers from the fact that the problem itself is connected with and embedded in a much larger array of issues. And this isn't seen by object thinking. The "solution" could bite back in a variety of ways, for example, by strengthening a tendency toward a one-sided diet in the illusion that "do-it-all" golden rice will prevent blindness. Or the genetic manipulation itself could affect other characteristics of rice in unforeseen ways.

Educating to Disconnect

All of us who have gone through conventional American schools have rich experience of the dominance of tests, especially multiple-choice tests, and grades. When you take a multiple-choice test the emphasis is on memorized facts, and what you remember either fits the choices or it doesn't. It's a clear world of right and wrong answers. Much cramming and little thinking is involved. Such an educational system is focused on and built around what British educator Stephen Sterling (2004) calls transmissive education. The goal is to transmit information from a source (a teacher, PowerPoint presentation, or a book) to a receptacle (the student). It has the "advantage" of being a neat and clean process; there are few vagaries. Everyone knows where they stand at the end of the day, since the test score tells all.

Transmissive learning focuses almost solely on content—the information to be learned or the knowledge to be had. In its most basic, stripped-down form (as expressed in multiple-choice tests), it is about exercising the capacity to mentally retain information. At its worst, this kind of learning is largely de-contextualized. Students may have the correct answer to the test question "What is a primary cause of global warming" by checking the box "carbon dioxide." But does this mean they have understood anything? Is this answer the end result of explorations into the biological, atmospheric, and social contexts of climate change? Or is it an isolated bit of information?

As the student may amass (and may proceed to forget) more or less disconnected packets of information, the school schedule enhances disjointedness by moving the student through seven forty-five-minute classes per day, each having little relation to one another in method or content. In each class there is a different teacher and a different learning group.

What is taken-for-granted in this way of educating—and therefore not recognized as an issue—is that when we teach students in a disconnected and disjointed way and, in addition, habituate students to see learning as culminating in test scores and grades, we create layer after layer of separation between the student and the world. Is it a wonder that they become alienated and lose interest?

Nonetheless, I was stunned when a supervising teacher from a state school in Germany, who had visited a class of mine at a Waldorf school to see if we were working on a requisite level for the state exam, stated, "your students are still interested in the subject!" He was surprised. Evidently, interest was something special. He then proceeded to tell me what I should change so that I would conform better to the state standards. I could hardly hold back paraphrasing his recommendations: you mean, I should implement these things so that I succeed in driving out the students' interest?

Although hardly anyone today, at least in theory, would deny the value of inquiry-based and experiential education over rote learning, our educational system has remained remarkably resistant to change. Symptomatically, the editor of *Science* magazine bemoaned not long ago the "typical school experience in which students memorize the names of plants and their parts from pictures in a textbook, often without encountering the actual object." He could only suggest that the way for the United States to have a "more engaging education system" is to develop after-school programs (Alberts, 2010, p. 427). In other words, given a cemented education system, experiential and exploratory learning will just have to occur outside of school.

If there were a god of object thinking who was pondering how to design an educational system to further his will—to habituate people to focus on pieces rather than wholes and on discrete facts rather than processes, and to see disconnection and separation as fundamental—he could hardly do better than what the American school system has already created.

Losing the Experienced World

The danger of any powerful worldview is that it loses sight of the fact that it does not encompass the totality of human experience. Our experience of the world we live in is clearly not exhausted by object thinking. When my son comes down the stairs in the morning, his dog stands up and wags her tail. He cuddles up with her and spends some time stroking her. This is not a world of interacting objects; this is a world of ensouled beings and is part of our everyday experience. Think of the conference room filled with tension, or the smiles on the faces of friends at the dinner table making a toast, or even the other driver who shakes his fist at you because you almost pulled out in front of him on the road. There may be objects in these interactions, but there are certainly not only objects.

There is no way that object thinking can deal with these kinds of experiences. It can ignore them, try to manipulate them, or deem them to be unreal. What is in any case clear is that beings, emotions, thoughts, and gestures have no place in a world of mere objects. They are "homeless" in a world of interacting atoms and genes, and they have no place in the space of objects. But they do not therefore cease to exist. Once, however, object thinking takes hold—during the scientific revolution and in early modern philosophy—thoughts and feelings become relegated to a private "subjective" realm that is separate from the objective world of things and to the essentially mechanical laws of nature. Science became a human endeavor only meant to deal with the objective—and a world devoid of life, soul, or spirit.

It is one matter to investigate the molecular and neural correlates of experience. It is another matter to claim that the experiences themselves are nothing real in themselves and that reality is confined to the robust, object-like ideas of science. The scientific literature—and popular science writing as well—is full of examples showing that scientists and editors conflate psychological phenomena (such as reason) and physical phenomena (such as neurons), as the title of an article in *Nature* illustrates: "Probabilistic reasoning by neurons" (Yang & Shadlen, 2007). The tragic thing is that many people don't even realize the absurdity of such statements.

The philosopher Whitehead spoke of the fallacy of misplaced concreteness in describing this form of self-forgetfulness (1967, pp. 51 ff.). It is essentially an unconscious process of reification—taking the knowledge one has gained through object thinking and then projecting that knowledge into the world as the fundamental reality. When we are asleep to the fact that the abstract concepts produced by object thinking are a result of human activity, we fall prey to the fallacy of misplaced concreteness. We take our abstractions more seriously and begin to feel them to be "more real" than the rest of our experience. Magically, in the process of reifying, the fullness of human experience is somehow supposed to disappear along the way. Or, at best, spirit and inwardness become a kind of numinous fog, an unnecessary "ghost in the machine" (Ryle, 1949, chapter 1).

It is a remarkable fact that many of the entities students are asked to picture—atoms, molecules, genes, hormones, neural transmitters, or even money—show themselves to be anything other than discrete things when you go more deeply into them. For example, quantum physicist Werner Heisenberg wrote that the atom of modern physics "is in essence not a material formation in space and time, but only, to a degree, a symbol, which, when introduced, makes natural laws assume a very simple form" (1947, p. 97). In a different context he writes, "In the experiments about atomic events we have to do with things and facts, with phenomena that are just as real as any phenomena in daily life. But the atoms or the elementary particles themselves are not as real; they form a world of potentialities or possibilities rather than one of things or facts" (1962, p. 186). As a result, in the words of another quantum physicist, Walter Heitler, "We do wrong when we want to teach young people something they can't possibly understand, or to misrepresent it in order to make it comprehensible.... I don't believe it is a good thing to talk about atomic physics and electrons in the upper elementary school. Every spatial representation of these formations is simply false" (1973, pp. 252, 256). That's a telling formulation: we misrepresent something to make it comprehensible. The misrepresentation often involves, for the sake of simplicity and in order to adapt the phenomena to one's own habit of mind, conceiving of something in a thing-like way that is in fact not so. This kind of misrepresentation is rampant in our culture and shows the force of object thinking.

We have to realize that an isolated fact or piece of information is, in fact, an artifact of the human mind that focuses solely on a particular aspect of experience. Even a rock you can take in your hand and study just by itself is in reality more than that object, since it is, for example, embedded in the earth's field of gravity and has a history— having at some point in time separated off from the bedrock in which it had its origin. Every single thing is embedded in the larger fabric of the world. Once we realize this fundamental truth, the tragedy of de-contextualized learning becomes all too apparent. We are training our children and students to look at the world as if it consisted of bits of information and separate facts that somehow have to be put together into a coherent picture, instead of letting them experientially participate in and understand, from the very outset, the interconnected nature of things—that to be a thing is to be a focal point of relations with the world. Philosopher Albert Borgmann describes this contextual nature of things in relation to a human-made object:

A stove used to furnish more than mere warmth. It was a focus, a hearth, a place that gathered the work and leisure of a family and gave the house a center. Its coldness marked the morning, and the spreading of its warmth the beginning of the day. It assigned to the different family members tasks that defined their place in the household.... It provided for the entire family a regular and bodily engagement with the rhythm of the seasons that was woven together of the threat of cold and solace of warmth, the smell of wood smoke, the exertion of sawing and of carrying, the teaching of skills, and the fidelity to daily tasks. (1984, pp. 41–42)

How little our culture values this rich world of meaning that shows itself in the manifold relations a thing can disclose. Instead, compelled by object thinking, we concentrate our attention on things as if they were independent entities.

I would argue that the roots of our ecological crisis—why sustainability has become a burning issue—lie in the way human culture has been driven by object thinking. From nature-as-object we can distance ourselves. With no "inside," nature becomes a "thing" that we can all too easily exploit. For the technological mind nature becomes exploitable resources:

When we look at a tree accordingly, we see so much lumber or cellulose fiber; the needles, branches, the bark, and the roots are waste. Rock is 5 percent metal and the rest is spoils. An animal is seen as a machine that produces so much meat. Whichever of its functions fails to serve that purpose is indifferent or bothersome. (Borgmann, 1984, p. 192)

To speak of nature's "intrinsic value" or "inherent integrity" is, for object thinking, only to project our own subjectivity onto nature—nothing real. In the end, why should we care about a bunch of things? The red of the cardinal flower is "only" a particular light wave, and water is "only" H_2O. Your feelings are "only" your hormones busily at work and your caring at best neuronal responses to stimuli. Why, in the long run, should we take interest in a world that is "only"? What moral commitment can I have to light waves, molecules, and hormones?

You may object that none of us consider nature in such stripped-down and callous terms. That is probably true. As individuals most of us are influenced by object thinking in less conscious ways. When object thinking becomes part of us, it can imbue us with a subtle but pervasive sense of alienation and distance to the world. It can give us false hopes that the solutions to complex problems can simply be implemented and everything will get better. Of course we are not object thinkers all the time, but we are very good in day-to-day life in using object thinking to compartmentalize: We may treat our pet dog as a precious ensouled creature at home and carry out animal experiments in the lab. In many ways it is almost impossible to navigate in our world today without exercising some kind of compartmentalization.

Regardless of how as individuals we find our way, there can be no doubt that in day-to-day life, in the practice of science, in educational systems, in the implementation of technology, and in the realization of capitalism around the globe *we effectively incarnate the object view of the world*. Object thinking goes to work in the service of a very powerful contingent of psychological characteristics, which can include those we often consider in negative terms (greed, ambition, fear, ruthlessness, conniving, etc.) as well as those we hold to be more lofty (compassion, idealism, care, love, etc.). There is no question that the object thinker in us can be motivated by the best of intentions. Paradoxically, it is precisely such psychological characteristics that drive object thinking, while object thinking has—in theory—relegated those very characteristics to an unsubstantial realm of epiphenomena. Object thinking creates an unbridgeable chasm between us and the world we set out to understand. At the same time, the world becomes an open field for manipulation and exploitation.

Beyond Object Thinking?

Systems theory or systems thinking is often viewed as an antidote to a fragmented way of thinking about the world, since it "looks at the whole, and the parts, and the connections between the parts, studying the whole in order to understand the parts. It is the opposite of reductionism, the idea that something is simply the sum of its parts" (O'Conner and McDermott, 1997, p. 2). But ecologist Frank Golley, in a fine book concerned with environmental literacy, points out the clear limitations of the systems approach he uses as an ecologist:

A systems approach is not a perfect vehicle for our purpose. The main criticism is that it is mechanical and treats nature as a machine. This is a serious objection—nature is above all else not a machine!—but in science I know of no better way to study and discuss wholes than in systems language…. In science we face a poverty of tools for exploring the nature of whole systems. (1998, p. 1)

This is a remarkable admission, which Golley reinforces at the end of the book where he asks, "How does one speak about connection in a culture of separation and isolation? I don't know" (1998, p. 231). Golley is acutely aware of the limitations of object thinking that are manifest in the scientific approach and recognizes these limitations even in the systems approach as applied within ecology, which is held to be the most holistic of sciences. But that doesn't mean it has transcended object thinking.

Philosopher Ken Wilber elucidates the essential limitations of systems theory: "Systems theory—precisely in its claim and desire to cover *all* systems—necessarily covers the least common denominator, and thus nothing gets into systems theory that, to borrow a line from Swift, does not also cover the weakest noodle" (2000, p. 122). Here Wilber points to the problem of any abstraction with which we try to "explain" phenomena. Abstract concepts or general theories may illuminate one aspect of a phenomenon, and the more fundamental that aspect is, the less it says about the concrete phenomenon itself. Thus the law of gravity may be applied to all bodies, but it tells us little about the difference between bodies such as cats, oranges, tape dispensers, and pebbles (Talbott, 2004b and Kauffman, 2008). For this reason Wilber concludes that while basic tenets of any given systems theory may be "fundamental," they are the "*least* interesting, least significant" precisely because they do not lead into the full, unique, and concrete reality of a particular phenomenon (p. 122; his emphasis). Basically, systems theory covers the external side of things:

The general systems sciences seek to be empirical, or based on sensory evidence (or its extensions). And thus they are interested in how cells are taken up into complex organisms, and how organisms are parts of ecological environments, and so on—all of which you can *see*, and thus all of which you can investigate empirically.... But they are not interested in—because their empirical methods do not cover—how sensations are taken up into perceptions, and perceptions give way to impulses and emotions, and emotions break forth into images, and images expand into symbols.... The empirical systems sciences cover all the outward forms of all that, and cover it very well; they simply miss, and leave out entirely, the *inside* of all of that.... And the empirical systems sciences or ecological sciences, even though they claim to be holistic, in fact cover exactly and only one half of the Kosmos. And that is especially what is so partial about the web-of-life theories: they indeed see fields within fields within fields, but they are really only surfaces within surfaces within yet still other surfaces—they see only the exterior half of reality. (Wilber, 2000, pp. 113–114)

Wilber speaks of this way of viewing—which leaves out interiority—as "subtle reductionism" (pp. 135ff.) and, as such, "flatland holism" (p. 136): "The systems theorists like to claim that the reductionist villains are the atomists, and that in emphasizing the wholistic nature of systems within systems, they themselves have overcome reductionism, and that they are therefore in a position to help 'heal the planet.' Whereas all they have actually done is use a subtle reduction to overcome a gross one" (pp. 136–137).[2]

This criticism is important. But it does not mean that systems thinking is of no value. Although rooted in object thinking, it can help us to become more aware of dynamic interactions and to expect that seemingly small changes can elicit large alterations. It can be valuable, for example, to analyze complex phenomena into feedback loops and thereby heighten our sense of reciprocal causation. Systems thinking can help loosen up the corset of object thinking that constricts the mind. But we need to go further.

In systems thinking it is as if we catch a glimpse of a different way of conceiving reality, but rooted in object thinking, we can only interpret and formulate what we've glimpsed in the very thought forms we need to transcend. Since object thinking is the foundation of our consciousness and gives us a sense of stability, it is not easy to let it go. It's like leaving solid ground and learning to swim. You have to enter the water if you are going to learn to move in it. The new medium demands new capacities, and those capacities can develop only by entering into and interacting with that medium.

Albert Einstein said soon after the atom bomb was dropped on Japan: "The war is won, but the peace is not…. Past thinking and methods did not prevent wars. Future thinking must" (cited in Calaprice, 2005, pp. 169–170). This statement has evidently morphed into a quotation attributed to Einstein: "We cannot solve problems by using the same kind of thinking we used when we created them." While Einstein did not say this,* the sentence captures the spirit of what he meant and expresses the point I want to make: Object thinking can stop a war and object thinking may even be able to reduce pollution. But just as the intelligence of object thinking has not brought peace to the world since 1945, it will also be unable to generate the flexibility of mind and the capacities we need to create long-term sustainability.

So the simple question that addresses a formidable task is: how do we move beyond object thinking?

From Object Thinking to Living Thinking

Through the foregoing characterization of object thinking we can already recognize some capacities we need to develop in order to move beyond its confines.

First, our thinking needs to become more participatory, rather than distancing and objectifying. We can begin by becoming acutely

---

* I asked Alice Calaprice, editor of *The New Quotable Einstein* (2005), if she could find this quote. She said it does not exist in that form. She believes the source of this statement is what Einstein wrote about war and peace, which I quoted above.

aware of the tendency to reify and fall in love with our abstractions at the cost of our attentiveness to the reality we actually live in. I am not saying we should lose, deny, or shun the ability to pull back and consider the world in quiet reflection. Rather, we need to wake up to the fact that, however we think, we are participants in the world. While the idea of a world out there, separate and independent from us, may be highly suggestive and comforting, it is an illusion. We need to realize how strongly we are bound up with the world, articulate the nature of our participation, and then infuse this awareness into our inquiries and educational practices.

Second, we need to become more concrete and less abstract in our thinking. Object thinking forms generalizations and abstracts from particulars. We need to counterbalance that movement of thought by a movement toward the manifold richness of the concrete world given to us in experience. We need to learn to move in the world of particulars in a way that allows us to disclose their essential characteristics. When we stop seeking material causes behind the phenomena, we can learn to see how the phenomena themselves are our teachers (Goethe 1995, p. 307) and heed phenomenologist Edmund Husserl's call to return "to the things themselves" (Husserl, 1993, p. 6)—and the "things" he means are not the things of object thinking; rather, he is appealing to a return to the immediacy of experience.

Third, we need to begin, in all earnestness, to orient our thinking around the living organism and living processes instead of around the idea of interacting object-like entities. The object view of the world has reigned long enough. We need a life view of the world. The animate, and not the inanimate, should become the main teacher for our thinking. We need an understanding of the living world through which the qualities of life become part of our own thinking, so that when we interact with each other, other creatures, and the rest of nature, our way of thinking is informed with the characteristics that life itself exhibits.

What are the characteristics of an organism from which we can learn? You and I are organisms. When I'm walking and get caught in a downpour, I start running. Running is not simply an isolated activity of a part of the body, with everything else staying the same. No, everything in my body moves into the running mode. Not only my muscle activity changes. This shift in muscle activity is in itself dependent on my attention, abilities, and the work of the rest of my body. All facets of my circulatory system—heart rate, blood viscosity, gas exchange, diameter of the peripheral vessels, and more—are working in the running mode. Running also shows itself in my digestive organs, in the liver, in the kidneys, and in my brain. You would be hard-pressed to find something that was not affected by my running.

And then I stop, catch my breath, and sit down at my desk. I gaze out the window at the pouring rain and the dark sky. A thought arises in my mind—I had better get to work now. Physically I'm here, but it's time to focus on the papers lying on my desk. Now I'm in think, write, and sit mode. Again, everything is shifting, working in service of what I want to accomplish.

This example illustrates a key feature of all organisms—their active existence as integrated functioning wholes. In an organism there is a very particular relation between the parts and the whole. Philosopher Immanuel Kant describes this relation succinctly: "Every part not only exists by means of the other parts, but is thought as existing *for the sake of* the others and the whole—that is as an (organic) instrument.... Its parts are all organs reciprocally producing one another.... It is an *organized* and *self-organizing being*" (1951, §65, pp. 220–1; emphasis in the original). Or in Lewis Mumford's words:

> To preserve wholeness in the midst of constant change, and to allow for a maximum amount of instability and variability, for adventurous effort, pushing beyond immediate needs and stimuli while retaining a sufficiently constant structure and a dynamic pattern of wholeness defines the nature of living organisms as opposed to random samples of molecules. (1970, p. 397)

These characterizations point to essential qualities of an organism: (1) the wholeness that informs an organism in each and every part and function; (2) the interconnectedness and reciprocal interaction of all parts and functions; and (3) the dynamic adaptability of an organism.

A thinking modeled after living organisms would be as dynamic, coherent, and responsive as a living organism.

Organisms do not exist in bubbles. Their boundaries are interfaces, allowing them to live in continual interaction with the environment that is the basis of their existence. A thinking modeled after life would be a thinking that is relational, that recognizes how living "things" interpenetrate and, in reality, are not things at all. It would be eminently context sensitive. It would be flexible enough to understand transformation, and it would be prepared for surprises. It would also be modest.

The cultivation of this kind of thinking is what should be at the heart of education. In his book *Earth in Mind*, David Orr speaks forcefully of the dangers of our modern educational system and also of the need for radical transformation:

> It is time, I believe, for … a general rethinking of the process and substance of education at all levels, beginning with the admission that much of what has gone wrong with the world is the result of education that alienates us from life in the name of human domination, fragments instead of unifies, overemphasizes success and careers, separates feeling from intellect and the practical from the theoretical, and unleashes on the world minds ignorant of their own ignorance. (2004, p. 17)

Orr is describing qualities that inform education—alienation, domination, fragmentation, goal-drivenness, separation of heart from head and the head from the hand, and arrogance. The characteristics needed for true sustainability education, as Stephen Sterling describes them, are starkly different: "flexibility, resilience, creativity, participative skills, competence, material restraint and a sense of responsibility and transpersonal ethics to handle transition and provide mutual support" (2004, p. 22). These are, essentially, organismic characteristics.

It is relatively easy to talk in general terms about all these organismic characteristics and about a participatory, holistic thinking. It is a wholly other matter to find concrete ways to develop living thinking—no small task in a culture so deeply encased in object thinking. But the plant can help us.

CHAPTER 2

Rooted in the World

Do you seek the highest, the greatest?
The plant can be your teacher:
what it is without volition
you can be willfully—that's it!

Friedrich Schiller

WHEN AN ACORN FALLS onto the ground in the autumn it comes to rest in a particular location. It may be eaten soon thereafter by a mouse. It may rot in the autumn rains. A squirrel might pick it up and carry it in its cheek to another part of the woods, dig a hole, and place it there. Even in this case the acorn's fate is still open—whether the squirrel digs it out and feeds on it in the winter, whether it decomposes, or whether it germinates and grows into an oak sapling.

Before germination, the life of the plant is encapsulated in the protective sheath of the seed (and in many cases, of the fruit as well). In this stage life is held back—full of possibilities yet to be realized—until the seed gives up its encapsulated state and opens itself to the environment. The opening often has preconditions: there are seeds that need a period of dormancy before they will germinate; others need to germinate soon after separation from the mother plant, otherwise they die; some seeds need to go through a period of cold before germination, while others even need to experience extreme heat (fire) to allow them to germinate. Whatever the specific and intriguing prerequisites may be for germination, the movement from the state of encapsulation to the actual unfolding and development of the seedling is a significant moment in the life of the plant. The plant's life can only unfold when it gives up being an object, when it grows out into and connects with the world in such a way that the world supports its further development. It cannot be a plant—which means to be a becoming being—unless it gives up its isolation and draws from the world.

Seeds are the most compact, solid, and, from an external perspective, the most self-enclosed, object-like stage in the life of the plant. Seeds are drier than other plant parts, and a key moment in the opening to the environment occurs when the seed casing allows water to penetrate into the seed, tissue swells, and the casing breaks open. The seed thereby forms a connection and continuity with the fluid environment. The water also allows its physiology to become active—what was solid as stored nutrients becomes fluid, and growth begins. Since water is the medium of active life processes, it is perhaps not so surprising that the generative (meristematic) tissues of the plant consist of 80 to 90 percent water; even wood consists of about 50 percent water. (On average, only around 2 percent of the live weight of a plant consists of what was taken as dissolved minerals from the soil.)

Regardless of the position in which the seed finds itself in or on the soil, when it germinates the seedling begins to orient itself in the environment: the root grows downward into the soil and the shoot grows in the opposite direction, away from the earth, and into the light and air. In growing straight downward, a primary root orients toward the center of the earth. We can imagine the taproots of all the plants on the planet as growing toward this center. So when the plant develops one pole in its roots that grow into the earth and another pole in its shoot that grows away from earth, it is placing itself into a huge planetary context. But it is also and importantly relating to its immediate, concrete environment. Whether the seed germinates at all and how it develops depend on what it meets when growing out into the environment with its particular and ever-changing constellation of light, wind, moisture, animal life, soil consistency and chemistry, etc. As plant ecologist Walter Larcher remarks, "the process of emergence and the seedling stage represent a particularly sensitive period" in the life of the plant (2003, p. 312).

The foremost activity in early development is rooting—the plant connects with and anchors itself in the soil. The root of the bur oak seedling grows rapidly into the soil (see Figure 2.1).

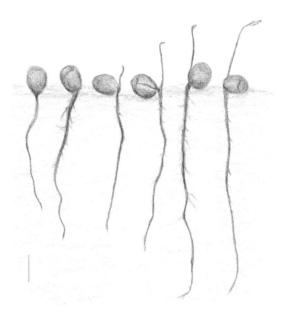

Fig. 2.1. Bur oak (*Quercus macrocarpa*) seedlings, showing development following germination; scale bar = 1 inch (after Holch, 1931, p. 268).

Shoot growth follows. Root growth draws from the reserves of the past season that have been stored as nutrients in the seed. It is important, when trying to picture growing roots, to realize that roots grow near their tips and that roots grow throughout the life of the plant. The primary downward growth of the primary root is initiated immediately behind a protective cap at the tip, and the same is the case for the lateral roots that develop over time. So in imagining the development of the rooting body we have to picture generative activity at the periphery, in all the root tips. Just behind the tips, roots develop fine root hairs that are the active interface with the environment. They increase the surface of the roots immensely and take in water and dissolved minerals. In this way the plant establishes intimate contact with its soil environment. Most plants not only open themselves to interaction with the soil directly but also join together with fungi to form a symbiosis that extends the plant's life even farther into the soil environment. Through these mycorrhizal fungi, the roots' absorbing surface for water and minerals is increased significantly and in return the fungi receive organic nutrients from the plant.

The roots are not only active in growth and taking up moisture and minerals, they also secrete substances such as acids into the soil, an activity that chemically alters the soil and allows the plant to access minerals it would otherwise simply pass by.

In growing upward into the light and air, the shoot-pole of the plant opens itself in a different way to different qualities of the environment. In contrast to the dense medium into which a plant roots itself, the shoot grows upward into the more rarified environment of light and air. In so doing it forms leaves that spread out as surfaces into this environment. Through its leaves the plant bathes itself in the light and air. The leaves have tiny pores—usually on the underside—through which air enters and departs. The air circulates through air-filled spaces in the leaves and becomes part of the plant's "food." In the presence of light the stems and leaves become green and in greening they can utilize the light of the sun for the plant's growth and development. Through light, carbon dioxide from the air, and water and a small amount of dissolved minerals from the soil the plant builds up its own body.

We should take a moment to appreciate this remarkable capacity of the plant. The plant can make its own living substance on the basis of light, air, water, and small amounts of dissolved minerals. What a contrast to our animal way of life that demands we live from already existing plant or animal substance. How different it would be if, to have a meal, we could go out and expose ourselves to the sun for a number of hours while drinking lightly salted water! But that is not how we are organized; we are more enclosed from the immediate environment, while plants have "an open form through which the organism in all its manifestations of life is directly integrated into its environment" (Plessner, 1975, p. 219). By taking root in the earth plants become in a way more dependent on their environment and more vulnerable than a roaming, self-mobile animal. But this dependency is the flip side of openness to the environment and the plant's ability to engage with that environment and to do what animals cannot, namely create, essentially out of air and water, living substance.

Figure 2.2 shows three representative bur oak saplings that grew in three different environments within a quarter of a mile of one another in eastern Nebraska.

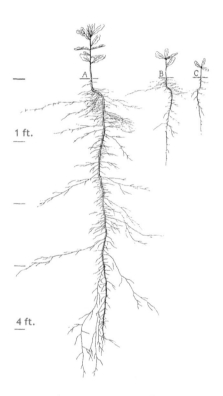

Fig. 2.2. Three different bur oak (*Quercus macrocarpa*) saplings growing in different environments, shown at the end of the first season of growth (after Holch, 1931, p. 274).

The soil was removed from the roots and the plants were drawn at the end of the first growing season. In all cases the soil was similar— "a fine silt loam known as loess" (Holch, 1931, p. 263). The plant on the left (A) was growing at the top of a hillside that had previously been cleared for cultivation and subsequently supported some prairie grasses; the acorn from which this seedling grew was planted in an area free of vegetation. It grew rapidly and deeply in this sunny environment with the rich, relatively dry prairie soil. At the end of the season the roots had penetrated the soil to a depth of 5 feet. Other bur oak acorns were planted nearby in a moister and shadier oak-hickory forest that spread out along a southwest facing slope (B). Here the tap root grew little more than a foot into the soil and formed proportionately fewer side roots. Finally, when the bur oak acorns germinated and grew in a darker, still moister linden (basswood) forest on a north-facing slope, they grew even more slowly and branched little (C). In all cases the above-ground part of the plants remained shorter than the rooting body. But above-ground growth was clearly correlated with root growth: the large-rooted plant also formed a longer main stem (what will become the trunk) with more leaves than the seedlings growing in the shadier, moister conditions.

What this example shows vividly is that by living its life through connecting with a specific place in the world the plant opens itself to the conditions of that place and interacts with them. Because the plant is an open, interactive being, the world it interacts with also becomes embodied in the plant's form and function. In opening itself to what comes to it from the environment and expanding out into that environment, it takes up an active relation to its surroundings, which then become the plant's environment. Place is not only the "location" that can be precisely defined in terms of longitude and latitude. Place for a plant is a web of relations that becomes manifest through the plant's life, substance, and form. The place-as-environment is what allows the plant—in a dynamic sense—to live; it is what the plant interacts with, it provides the plant with what it needs to live, and at the same time it is changed by the life of the plant.

Already in this brief consideration of plant germination and seedling development we see essential and intertwined qualities of plant life: how it embeds itself in a place; how it opens itself to the environment in which it grows; how it transforms itself as it develops from one state to a next while maintaining overall coherence of the organism; how plasticity allows it to develop in relation to particular environmental conditions; how it embodies the environment in its forms and functions; how it extends beyond itself as a bounded body (think of the mycorrhizal symbiosis) and is a member of a larger living context. All these themes will concern us throughout the following chapters, especially in chapters 3 and 4. Here I want to focus on how we can learn from the plant as a creature of place and from its remarkable openness to its environment in relation to the question: how can we as human beings develop a more living relation to the world?

Becoming Rooted — Perception

It seems as if the day was not wholly profane, in which we have given heed to some natural object. The fall of snowflakes in a still air, preserving to each crystal its perfect form; the blowing of sleet over a wide sheet of water, and over plains, the waving rye-field, the mimic waving of acres of houstonia, whose innumerable florets whiten and ripple before the eye; the reflections of trees and flowers in glassy lakes; the musical steaming odorous south wind, which converts all trees to windharps; the crackling and spurting of hemlock in the flames; or of pine logs, which yield glory to the walls and faces in the sitting-room,—these are the music and pictures of the most ancient religion. (Ralph Waldo Emerson, from his 1844 essay "Nature" [1990, p. 312])

In these descriptions Emerson shows us that he has "given heed" to the world around him. Actually, to say "around" him is not correct. In perceiving these occurrences he was out with them and took them in; he participated in them. Only then could he describe his experiences of nature so concretely as moving, unfolding processes. In such meetings with the sense world Emerson experienced something deep—the day is not "wholly profane," and he intimates an "ancient religion," a reconnecting with the roots of existence.

Most of us have experienced immediate and deeply enlivening meetings with the world—the smile of a young child; the rainbow arching across the light-bathed sky; the glowing red and orange clouds of a sunset; the waves building, breaking, crashing, and running up onto the beach. Such experiences are powerful and yet fleeting; we know ourselves as affected by them—we have met something and been nourished by something greater than ourselves. The experiences I've mentioned are special ones; they are not necessarily day-to-day occurrences. And yet, most of the waking day we are in the process of perceiving in some way or another.

But everyday experience becomes "merely everyday" and loses vibrancy inasmuch as it shrinks into intellectual thoughts, interpretations, biases, and categorizations of experience. Often we only notice something insofar as we already know it. I see a "dandelion," but how much of its radiant yellow do I really take in and acknowledge? I see the "pond," but I don't notice the undulating waves or the reflections of the trees and the sky quivering on its surface and extending into its depths. In one important way our experience is deadened because our perception has narrowed to what we already know. The world becomes prosaic, a world of things that is scarcely alive with the music of a resounding world.

A plant opens itself to its environment as a prerequisite for unfolding its life. It puts itself out into the environment. This openness to the environment does not end once it has germinated and established itself as a seedling. The roots continue to grow and near the tips remain in active interplay with the environment. The leaves spread out, new ones develop, and interaction with light and air do not cease. As vital organs the roots and leaves don't stop being open and close off from the environment, saying, physiologically, "We've had enough interaction." So the plant's openness to the environment entails initial receptivity, the activity of expanding out and ramifying into the environment, and the ability to remain receptive as it continues to interact with the environment.

These are also the fundamental gestures of human perception. When I am immersed in thought and then a pileated woodpecker hammers into a tree in the nearby woods, my attention is drawn out. I live for a moment in the sound and in its reverberation through the trees. In being in the sound I'm receptive. In fact, at that moment there is no "I am here" and "the bird is over there." There is simply the sounding in which I am participating. I am changed and grow richer through this experience.

I suggest that a prerequisite for gaining a living relation to the world as human beings is the ability to open ourselves through attentive perception. This living relation begins when we go out, actively and yet in the mode of receptivity, take in, and then engage with what we discover. In the process we become beings of place, even if we are on the move. We are attending to and taking in some of what the world offers up. In contrast, we are placeless when we are caught up with or consumed with ourselves, when we notice only what we have known before. If we want to open ourselves and root ourselves in the world in a living way we need to develop pathways to get out into experience, to become more conscious of immediate experience, and to learn to work with our ideas in such a way that they do not place barriers between ourselves and the richness of the world.

So a key issue is: how can we become more open and remain open to the richness of the world? Can we learn from the plant a way of being and, to paraphrase Schiller, do willfully what it does organically? This demands a kind of active wakefulness on our part to "be there." Or we could say, developing presence of mind as a kind of peripheral attentiveness, a readiness to take in. This is no simple matter and certainly, for me, not a given. It is a skill to be developed. In what follows I will describe a number of different ways "to get there from here," most of which are based on adult education courses at The Nature Institute.

Into the Phenomena

Many weeklong summer courses at The Nature Institute include plant study. One of the first observational exercises we carry out is the following. We go outside and I ask everyone—twelve to twenty participants—to look at a particular species of plant. I have selected the plant beforehand: one that is flowering and can be easily found in fields or along roadside edges. We walk around and see where it is growing. I ask everyone to take a few minutes, look at the plant and its surroundings, and then pick one specimen to bring inside. Back inside, we sit in a circle, each person with his or her plant. I give some guidelines for our observational process: We will go around the circle and each person will describe an observation of the plant. I request that descriptions be kept fairly brief, so that everyone gets a chance to share observations with the others. I ask that we try not to repeat what others have said, a suggestion that encourages mutual listening. I also request that those participants who may know botanical terms use them only if everyone else can follow the description. Finally, I say that we are not concerned here with explanations, causes, or models. We are not asking "why" questions; we simply want to take in and describe what the plant has to offer.

So we describe, moving from the bottom to the top of the plant. I will not try to reconstruct the whole process, but just give a few examples. A person is looking at the lower part of the stem and describes the clear transition between the whitish root stalk and the upright stem, which at its base is purplish and then turns green. Someone else describes the stoutness of the stem and the fine hairs that are mainly present along the slight vertical ridges running along it. Another person describes the oval shape of the lower leaves with their basically smooth margin, and notes the marked veins, especially visible on the leaf's underside. You can imagine that with such detailed observations and descriptions, we are carefully attending to what can be seen, felt, and smelled on the plant. We might go around the circle two or three times until we have a sense that we've attended to the different features of the plant. Such a process takes at least an hour if not more. Sometimes it will be continued the next day.

Although deceivingly simple, this process yields many fruits. First, and perhaps foremost, it is a cathartic practice to step out of everyday habits and to simply give one's full attention and time to something one would normally, at best, take in only at a glance. It helps us realize that we almost never look at things in a careful and detailed way. How often we gloss over things! Moreover, we are impressed by the plant in all its detail, pattern, and variability. In one course we studied common milkweed (see also chapter 5) and one person wrote in her evaluation: "I always look at milkweed differently now. I had the profound experience that, even as a total novice in the life sciences, I could, through attentiveness to the natural world around me, come to know it better." This can happen with the most inconspicuous weed. So by looking carefully we take the plant seriously—we turn our unencumbered attention toward it. We see the plant as something in its own right and learn to value it for its own sake. As one course participant remarked, "I will never walk past a daisy the same way!"

If we were to look at the plant from too narrow a perspective, this realization might well not occur. If we were interested only in, say, what medicinal properties a plant has, we could get a quick answer from an expert or book. But we are not carrying out a question-and-answer session with the plant. Instead, we are taking the time to perceive, to dwell with the plant and its features.

In this exercise we also notice that there is no natural end to observing. There is—even if we don't dissect, use microscopes or do biochemical analyses—always something more to see, smell, or touch. In this sense, the perceptual world has endless richness of detail and pattern to disclose. It's only we who choose to stop perceiving at some point. For most people this discovery is a kind of "aha" experience. We get a glimpse of what philosopher Merleau-Ponty calls the "hidden and inexhaustible richness" of the sense world (1969, p.139). Reflecting back on a weeklong course, a participant wrote: "Now I understand that the course is really about us, *Homo pretentious*, and the plants are what we work with because they're accessible, compliant, free and easy, and yet perfectly capable of revealing Nature in full glory to all who care to look. One of my chief impressions of the week is that almost any small bit of Nature will do the job."

Something else is remarkable in the process of group observation. We notice how differently people perceive and describe. Everyone in the circle realizes that, alone, he or she would not have seen nearly as much. Our senses are opened and directed in new ways by what others perceive and comment on. Some people have an ability to see more and more within a detail others don't attend to, like the participant who never left the root, even after we'd gone around the circle four times. Or the person who noticed the different shades of green, or how the plant felt when she waved it back and forth as if in the wind. So the plant reveals more and more of itself as different people make different discoveries. Knowledge arises in a community. Through such a process a learning community develops, and, in Goethe's words, "The interest of many focused on a single point can produce excellent results" (1995, p. 12). The unique perspective each person takes truly enriches the whole.

What allows different perspectives to show their best sides is the fact that everyone's attention is on a phenomenon that people don't have a great deal of pre-knowledge (prejudices and assumptions) about. They can look in quite an open way. Even people who have studied botany have rarely looked at one plant for so long and in such detail. Also, it's not about what we know from memory or our book learning, but about what we perceive *right now*.

Different people can have different perceptions, but these differences do not create separation; they enhance one another. We learn to appreciate the different ways people observe and describe. There may at times be need for clarification and more precise or accurate formulation, but that can all be achieved through recurring attentiveness to the thing itself and through mutual struggle to find ways to adequately express what we've perceived. The plant is a natural corrective for flights of fantasy or mere opinions. All we need to say is, "Look again, how is it really?"

This kind of observation exercise takes us into details and we attend to what is directly before us. A focus is chosen and is clearly circumscribed. To repeat the remark of a course participant, "Almost any small bit of nature will do the job." Although the chosen focus I've described is plants, we could (and sometimes do in our courses) immerse ourselves in a rock, a section of a meadow, a cloud formation, or an insect. Everywhere we focus our attention on the natural world we will discover an "inexhaustible richness."

Nature Drawing

Drawing can help facilitate looking. As John Ruskin noted in his classic *The Elements of Drawing*, "We always suppose that we *see* what we only know" (1971, p. 28). We all "know" that a blade of grass is green. We may even believe we see it as green when, in fact—if we put aside our preconception and actually look—the blade of grass is yellow in the particular light conditions in which we are observing it. Anyone who looks closely observes that color is dependent on the illumination. We have to look; we can't "know" the color beforehand. Similarly, we may know that the form of a building is rectangular, but when we attend to what we see, from the particular standpoint we have, we notice that if we draw a rectangle for the face of the building we are drawing something that looks completely wrong. So drawing can lead us out of our mental preconceptions and into the appearing phenomena themselves.

Ruskin spoke of regaining a childlike "innocence of the eye" (p. 27) that can open our perceptions and give us the possibility to draw what we see: "For I am nearly convinced that, when once we see keenly enough, there is very little difficulty in drawing what we see…. I believe that the sight is a more important thing than the drawing; and I would rather teach drawing that my pupils may learn to love Nature, than teach the looking at Nature that they may learn to draw" (p. 13). Drawing in this sense is a schooling of seeing—a way of opening up our looking and orienting it around the fine nuances of form and color. One course participant remarked: "Of most value was the increasing ability to see and to see how little I see. I feel that my eyes have been newly enlivened and I want to keep drawing."

In some courses we started by drawing a white ball that was lying on a cloth; the ball was illuminated from one side so that it threw a shadow onto the cloth (Figure 2.3). This setup provides a wide spectrum of light and dark shades and the "simple" elegance of the sphere. We draw in such a way that we do not make outlines—a line as a boundary is the creation of the intellect; what one sees are shades

Figure 2.3. Shading.

Figure 2.4. Ficus leaf.

Figure 2.5. An Oak leaf — two perspectives.

of light and dark. (That most of us tend to draw outlines shows the degree to which we are governed by object thinking.) In trying to put these shades on paper, we notice how the object emerges out of the interplay of light and dark and how its bodily, three-dimensional aspect becomes all the more "visible" on paper the more we can do justice to the seen patches of light and dark and the transitions between them.

On the basis of such an exercise we can turn to an organic form—a simple leaf from a *Ficus* bush that grows indoors as a house plant (Figure 2.4). Again, we attend to what we see, which allows the softly undulating form to appear.

A different drawing exercise leads us more into form qualities (Figure 2.5). Here the task is to take a leaf—in this case a white oak leaf—and sketch it first by filling out the form from the inside out. Again, no outlines but shading from the center and moving toward the edges. It's not so important that every detail is "right," but that the form emerges centrifugally out of the pencil marks. Then you do the exact opposite: you start by shading the outside—the space around the leaf—and move in toward the leaf margin. In this way you draw the negative space around the leaf, and the leaf emerges as the "empty" space in the middle. This trains our observation to attend to a form in relation to its surroundings. Drawing the leaf from the inside out is much easier and comes more naturally—natural for our object relation to the world. It is more difficult to take the space around the leaf to be "real," draw it, and let the leaf emerge in this way.

The process of drawing presents challenges, because you often struggle with your own limits in technical facility, but if that concern can be overcome, you make discoveries. In the words of a course participant:

> The drawing started out very difficult for me. I wanted to do it and be done. I was surprised as I sat with my drawing and "what" I was drawing, that I really could take the phenomenon in and express it on the paper by going into it. It helped to bring up and demonstrate the going in and going out. Working in one color only also was amazing in helping me feel subtle gradations rather than discrete "things."

Through drawing we are, literally, drawn out into the phenomena.

Sauntering of the Senses

Another kind of exercise complements this focused attention to detail in observation and drawing, an exercise that asks us to let our attention spread out and wait to find what comes toward us. Here we don't predetermine what we attend to but, in a sense, invite the world to speak. Thoreau describes the intention:

> I must walk more with free senses—It is as bad to study stars & clouds as flowers & stones—I must let my senses wander as my thoughts—my eyes see without looking.... Be not preoccupied with looking. Go not to the object, let it come to you.... What I need is not to look at all—but a true sauntering of the eye. (September 13, 1852; in Thoreau 1999, p. 46)

This is decidedly difficult. In a course called "In Dialogue with Nature," which took place on one Saturday per month for ten months, we took a walk each of the ten afternoons into a forest and wetland preserve. I said we would go out and, at first, simply attend to what caught our attention. I gave no further instructions for this part of the walk, which was definitely not easy for everyone. As one participant wrote in her evaluation, "I found this the most challenging part of the day—being with the elements of weather, and also noticing my observation skills and memory sadly lacking. I would welcome less time doing this." But others responded differently: "I really enjoyed experiencing the same place in all of the seasons and in all kinds of weather." What we all took with us from month to month was a sense of open expectation about what would be different each month. We couldn't know ahead of time; we had to go out with our legs and attentiveness and see what would come to meet us.

The intent of going for an unstructured walk (which is also not easy for the teacher!) is to encourage what Thoreau called "sauntering of the eye." The walks actually oscillated between the more open expectant perceiving and the shift to observing closely what caught our attention. On the walk in October we noticed the vibrant and radiant colors of the leaves that shone to us not from the tree tops but from the forest floor. The forest floor was alive with color from the leaves that had descended in the past days from the tree crowns. We picked up some of the leaves and looked at the variations in color—and also noted the tree species. In November, reaching the same spot in the path, the forest floor was now muted, having turned a more homogenous yellow-brown. On the October walk I also noticed that a small tree was flowering—radiant little yellow flowers. Since no one else called attention to these flowers, I did. Everyone looked and was intrigued: a tree that flowers in October, when everything else in nature seems to be receding. Each time we returned to this spot in the coming months we looked at the tree. The flowers faded and fruits began to form. Over these months we made acquaintance with witch hazel (*Hamamelis virginiana*).

It was clear that participants on the walk often needed me or others to help them get out into perception and to be receptive to what was there to be perceived. But precisely that experience is educative. It is much easier to describe a plant in detail than it is to go sauntering with open senses. In the latter case we must willfully try to open our attentiveness and invite the world in—we have little control, and that is both unsettling and cathartic.

One exercise that helps bridge the gap between controlled focus and the ability to saunter with the senses is to choose a broad sensory focus for attention. For example: I go for a walk and say to myself, "I will focus my attention today on colors." Or I will focus on smells, or on sounds. This focus by no means determines what you will see, smell, or hear, but by narrowing your attentiveness to a sensory modality you are more receptive to that realm of experience. One day I was walking in the above-mentioned forest and wetland preserve with the intent of paying attention to light in the forest. I began noticing what I otherwise took for granted and had not really seen at all: the dark areas, the spots that were very bright, the more diffuse columns of "sunbeams." The wind was blowing on that day, so there was an ongoing play of changing illumination. At one moment a spot lit up brightly, changed form, and disappeared. I was strongly struck by this appearance. I don't know why, and I cannot describe it any further. But it was a deep experience and one that I can remember back to, although that memory is by no means the same thing as the one-time, striking experience itself.

I apply such sauntering of the senses in my research. Let me give an example. I have studied the skunk cabbage (*Symplocarpus foetidus*; Holdrege, 2000b). It grows in the wetland area of the forest preserve and over many years I have visited the skunk cabbage colony. Sometimes I go there with a specific question or task in mind: Are the flowers out yet? What are some of the main differences between the small plants and the large plants? Can I find seeds? Other times I go out with a more open frame of mind—my attention is naturally drawn to skunk cabbage, but I am not focusing on any particular task or question.

One March afternoon I went down to the wetland in the mode of sauntering of the eye. The sun was shining through the leafless shrubs and it warmed my face. My eyes were wandering over the skunk cabbage flowers that were just emerging from the cool muck. Then I noticed a few honeybees. I watched those bees fly in and out of the bud-like leaves that enwrap the skunk cabbage flowers. In a flash I realized that I hadn't seen any bees yet that year. The first bees of the year were visiting this plant. What a wonderful meeting of bee and skunk cabbage that I had never seen before. I was in the right place at the right time. I had been going regularly each year to the wetland on early spring afternoons, and yet I hadn't seen this before. I'm pretty sure I may have continued to overlook the meeting of bee and skunk cabbage had I not been practicing a "sauntering of the eye." I would have been too preoccupied with other things on my mind that would have blocked my ability to be present with what was occurring around me. Having seen this event caused great joy, but it also sparked a small revolution in my understanding of skunk cabbage and bees.

I have described two complementary types of sensory observation exercises. In the one case—the example of the plant observation—we go out with our attention to meet something particular and take it in with all its details. We move with our senses and attention in and through the phenomena. In the other case, we try to create a kind of open receptivity that allows us to take in what appears at a given moment.

Every perception of a thing or situation has these two aspects—focus and receptivity. Without these there would be no perception. By carrying out such exercises in both directions, we are honing our capacities to perceive the world around us. In this way we can shift into a sensory mode, being with the things themselves. The two kinds of exercises actually reinforce one another.

One day I was struck by all the white feathery globes of fruiting dandelions that seemed to hover over the rich green grass. I then took some time to look at the globes more carefully. Each consists of an array of compact fruits still barely attached to the top of the stem, each of which sends out a fine filament that radiates out into the fine hairs that together form the circumference of the globe. The spheroid globes make a regular shape, although the individual fruit extensions are not all the same length; some are shorter, some are curved, and together they form the globe. A structure you can only be amazed at, and you realize: I have not begun to fathom the genius of this organism. Two days later we had—in the middle of the night—a violent storm with heavy rains and strong winds. The next morning I looked out the window and noticed that all the globes were gone—the storm had dispersed them into the wider world. How many times have I looked out a window in "dandelion season" and never noticed the disappearance of the globes?

I was attuned to dandelions by having taken some time to look at them carefully, and, enlivened by them, I did not overlook their subsequent transformation. So it happens: the world catches our attention although we are not attending to what draws us out of ourselves—I wasn't intending to look at dandelions when I crossed the yard and noticed the globes; the globes stimulated me and I looked more carefully and was even more amazed. By going into the phenomena in their details I awaken more to the world. Then I am more open to that part of the world which I have noticed and don't pass by it so easily.

Exact Sensorial Imagination

When we have made the effort to perceive carefully, this interaction leaves an impression on us. We can remember, at least to some degree, what we have seen, felt, smelled, touched, or heard. So after we have carried out a variety of observation exercises, I request that participants in courses willfully re-picture or re-create in their imagination what they have perceived. Picture the color of the stem and how it changes from bottom to top, feel the consistency of the stem by imagining the feeling of the pressure you applied to it with your fingers, reawaken the fragrance of the blossom and dwell in it for a moment. In this way we can build up a vivid picture of the plant we have observed or of the forest we have walked through. By doing so we awaken in ourselves what we have met through sensory engagement. We can actually remember much more than we realize, and, moreover, sometimes the hue of green or the shape of a leaf will speak more strongly in our inner picturing than it had in the moment of observing.

During a course I ask participants to make re-picturing into a daily practice: picture in the evening or morning the plant or environment with which we have been concerning ourselves. We talk about the experience of picturing, and people share their questions and approaches. It is fascinating how differently people picture. Often people notice that they couldn't picture something because they hadn't really looked at it.

What is the significance of inner re-picturing, which Goethe called "exact sensorial imagination" (1995, p. 46)? First, it is a practice that allows us to connect ourselves consciously and vividly with what we have experienced. We bring to awareness what would otherwise sink into a sea of potential memories. We willfully call up these experiences and enter into them with our picturing activity. This activity is imbued with feeling: not reactive feeling but feeling as a connecting agent, an inner sensorium for qualities.

As in perception we go out to things and invite them in, so in exact sensorial imagination we re-create and enliven within ourselves what we have met in experience. In this way we connect deeply with the world we meet in sensory experience. One course participant described how the work in a course created "lasting experiences of the plants through the practical observation and visualization exercises—I feel I have 'met' two plants, as many of my perceptions still live fresh in my imagination." We have taken the plant in and now we move it in us. Or, said differently, we come into inner movement by re-creating in imagination the qualities we have perceived. We can thereby become more aware of these qualities.

Second, this practice can help us to notice that we need to perceive more carefully if we are going to be in a position to faithfully re-create in ourselves a vivid image. This realization motivates a return to the phenomena.

Third, it is an aid to overcoming the tendency to think abstractly. In exact sensorial imagination we are using our mental capacities to get closer to the concrete sensory qualities. This contrasts starkly with an abstract frame of mind that uses concepts to explain and interpret what we perceive. Exact sensorial imagination lets our minds practice concreteness instead of abstract distancing.[1]

We can view the perception exercises described in the previous section and the practice of exact sensorial imagination as two polar practices that enhance each other. Both need to be practiced. By going out into perception and openly taking account of what the world offers we inform experience with the richness of the sensory world. Through exact sensorial imagination we connect these experiences with ourselves and at the same time become inwardly active. The world comes to life in us. We can practice a kind of pendulum swing between going out and bringing in and enlivening, going out again, bringing in and enlivening. My personal experience is that by doing this, both perception and picturing are enhanced. Through careful perception I participate in the phenomena. This gives me a wealth to re-picture. Through vivid re-picturing my attentiveness to the world is enhanced. I perceive vividly and more can be disclosed in any moment.

The following is a symbol of this oscillation. Imagine moving along a figure eight. Starting at the middle you swing out into the

lower part of the curve and around upward and back to the middle, swing down and out to the right, up and around back to the middle.

It is important to think of this oscillation dynamically as a movement.

Now we can imagine that during this continuous movement around the figure, our attention is directed outward as we move on

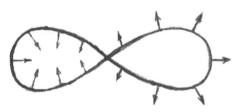

the right half of the figure, turns inward on the left half of the figure, and then turns outward again as we move along the curve. In perception we move out into the

world and then take what we have experienced into ourselves. Out of our own activity we can then enrich our experience of the outer (through, for example, exact sensorial imagination) and then move back outward. We are moving into the world and the world is moving into us. When we move far out into the world, we can move far into ourselves; this allows us, in turn, to expand more into the world. "Outer" and "inner" can no longer be viewed as two distinct realms; they are two aspects of one oscillating activity. We are, inasmuch as we bring forth this dynamic movement, interfaces in which we and the world continually intersect in vibrant activity.

Openness and Preconceptions

At present I am preoccupied with sense-impressions to which no book or picture can do justice. The truth is that, in putting my powers of observation to the test, I have found a new interest in life. How far will my scientific and general knowledge take me: Can I learn to look at things with clear, fresh eyes? How much can I take in at a single glance? Can the grooves of old mental habits be effaced? This is what I am trying to discover. (Goethe; September 11, 1786 [1982, p. 21])

As roots are stimulated in their growth through the soil conditions they encounter, so are we stimulated through our encounters with the perceptual world. Goethe "found a new interest in life" through putting his "powers of observation to the test." Since Goethe was an unusually sensitive and attentive observer, it seems at first odd that he needed a kind of challenge to stimulate his perception and to "look at things with clear, fresh eyes." At the time he entered these thoughts into his journal he was thirty-seven years old, a famous writer and poet. In addition to his literary creations, he had carried out an array of scientific studies in optics, geology, botany, and comparative anatomy. Moreover, during the previous ten years he had taken on more and more responsibilities in the small dukedom of Weimar. He was privy councilor, president of the ducal chamber, overseer of buildings and mines, president of the war council (the army had 660 soldiers), and director of the drawing academy (see Richards, 2002, p. 355f.; Barnes, 1999, p. 27). He had come to a point in his life where he felt stifled. He needed a change and decided that he must leave Weimar and gain fresh experiences. He hoped that the world would breathe new life into him. He decided to flee Weimar, incognito, and go to Italy. He remained there for two years.

He wrote the above words soon after he had crossed over the Alps and arrived in northern Italy. Freed from all outer responsibilities, he immersed himself in perceptions. For example, when he was traveling through the Alps by horse-driven coach he noted both new species of plants and species he was familiar with. But the familiar plants were different: "In the low-lying regions, branches and stems were strong and fleshy and leaves broad, but up here in the mountains, branches and stems became more delicate, buds were spaced at wider intervals and the leaves were lanceolate in shape. I observed this in a willow and in a gentian, which convinced me that it was not a question of different species" (September 8, 1786; Goethe, 1982, p. 15). He felt, as he put it, "a new elasticity of mind" (Goethe, 1982, p. 21).

The immersion in all the new experiences helped Goethe to efface the "grooves of old mental habits." And yet it is also clear that the willow and the gentian in the Alps spoke to him strongly because he had previously engaged in the study of plants. His knowledge of plants allowed him to see what others would have ignored, and at the same time his knowledge grew as it was illuminated by the encounter with a different situation.

The seed has developed out of the plant's activity and growth in the past year. When what has been stored in the seed becomes fluid, the plant can grow anew and enter into its environment. Likewise, in human life each present experience is informed by past experience. We are not unformed wax onto which the world imprints itself. We bring much with us into every sense experience and yet we do not want to be constrained by mental grooves. We want to be open to what we haven't noticed and to what we don't know. Thoreau states in his characteristically forceful way how our past learning and knowledge can encumber our perception:

> It is only when we forget all our learning that we begin to know. I do not get nearer by a hair's breadth to any natural object so long as I presume that I have an introduction to it from some learned man. To conceive of it with a total apprehension I must for the thousandth time approach it as something totally strange. If you would make acquaintance with the ferns you must forget your botany.... Your greatest success will be simply to perceive that such things are, and you will have no communication to make to the Royal Society. (October 4, 1859; in Thoreau 1999, p. 91)

Thoreau appeals to the ability to rid ourselves of preconceptions and to move into a state of "total apprehension," a state in which we can approach each thing as if we had never seen it before. But is he not asking us to do the impossible? If we achieved this state of ignorance in the face of each new experience, would we experience anything at all? Thoreau was not naïve. He was keenly aware that we need experience to see:

> It requires a different intention of the eye in the same locality to see different plants, as, for example, *Juncaceae* [rushes] or *Gramineae* [grasses] even; i.e., I find that when I am looking for the former, I do not see the latter in their midst.... A man sees only what concerns him. A botanist absorbed in the pursuit of grasses does not distinguish the grandest pasture oaks. He as it were tramples down oaks unwittingly in his walk. (September 9, 1858; in Thoreau 1999, p. 83)

In these two apparently contradictory statements Thoreau expresses the tension that exists in getting to know the world: on the one hand the striving for openness for which we have to leave our preconceptions behind us: to "make acquaintance with the ferns you must forget your botany;" and on the other hand the abilities of the trained observer that allow us to see more: "It requires a different intention of the eye in the same locality to see different plants." Without being open to see something new, we would see the world only through the lenses of our preformed conceptions and experience. But without any conceptions, we wouldn't see anything at all. Clearly, both Goethe and Thoreau lived creatively in this tension. And to live creatively in the tension you have to become aware of it, as both Goethe and Thoreau were.

That we open ourselves to new experiences is the basis of all learning, but our learning of the new is also always informed by what we have learned in the past. As philosopher Hans-Georg Gadamer describes it:

There is always a world already interpreted, already organized in its basic relations, into which experience steps as something new, upsetting what has led our expectations and undergoing re-organization itself in the upheaval.... Only the support of familiar and common understanding makes possible the venture into the alien, the lifting up of something out of the alien, and thus the broadening and enrichment of our own experience of the world. (1977, p. 15)

Gadamer calls this ongoing experience in life the "hermeneutical experience" (p. 15). It is what lets us continue to learn. Past experiences inform any new experiences. Our openness to the world will always have some kind of "initial directedness" (p. 9). But is this directedness elastic, flexible, and welcoming to the not-yet-known? Or is our inner direction narrow and rigid so that it impoverishes and makes all too one-sided our experience of the world? As Abraham Maslow remarks, "It is tempting, if the only tool you have is a hammer, to treat everything as if it were a nail" (1969, p. 15).

Becoming Aware of Thinking

In turning toward the world with the effort to take in and discover what it offers, we find that the world is always richer than what we can conceptualize. But it is also evident that much depends on how aware we are of the thoughts, feelings, and biases we bring to bear on every new experience. In chapter 1 we looked carefully at the objective attitude and at object thinking, and in this way became more aware of these thought styles that so dominate modern life. In this chapter we have considered the process of perception, which is always informed by our thinking. In concluding this chapter I describe an exercise we have practiced in Nature Institute courses that in a striking way helps people develop awareness for the fabric of mental life.

The fifteen course participants are sitting in a circle. As is typical for Nature Institute courses, about half of the participants are educators and the rest have a variety of professions: there are three elementary school teachers, a middle school teacher, two high school teachers, a biology professor, a retired geologist, two gardeners, a healing-plant specialist, an architect, a graduate student, an artist, and a businessman. We are near the beginning of a weeklong course called "Coming Alive to Nature."

I tell the group that I'm going to show them something they probably haven't seen before and that, in the time that follows, I'd like them to attend not only to the object, but also to what is going on in them—what are they doing while perceiving the object?

A paper bag lies at my feet, and I reach down into it and pull out the object. I raise it above the level of the tables—yes, it is smaller than a breadbox—and hold it up for all eyes to see. Some puzzled gazes, some smiles, a few gasps. After a minute or two I lower the object out of the visual field. I ask, "What should we do now? Can we leave it at that?" No. I can't simply pack away the object and move on to other concerns. It has awakened interest and curiosity. It is a strange object and doesn't simply fit into our everyday world. It is not "only" a new variety of toy or rock, it's a different type of thing altogether.

So I ask again: what should we do now? Most people want to get closer to the thing—literally. To look at it more closely, to touch it and smell it. So I pass it around the group with a few guidelines: don't let it fall; don't lick it; and don't try to alter its form by pressing it too hard (it should last for others to enjoy). Each person takes a minute or two to investigate the object. Meanwhile, the others watch and are mostly silent. (Readers of this book will have to come to The Nature Institute to get to know the strange object; I won't reveal it here, so as not to spoil anyone's future experience.)

People move it around in their hands. They stroke it. They move their hands and forearms up and down to gauge its weight. Some shake it to feel if something is inside. Some tap it gently to get a sense of its insides. It gets sniffed, often followed by a thoughtful expression or sometimes a frown. And so the strange thing wanders around the room and arrives back with me.

I ask: what now? People start asking questions. They want to know what it "is." It is clearly not enough for them to have seen, felt, and smelled the object. They know it a bit, but they want to know more. Is it natural? Is it human-made? We could easily start a question-answer session in the mode of "Twenty Questions." But I don't want to simply reveal what it is; there should be a process of discovery. So I ask the participants to imagine what they would do if I were not here to give guidance or answer questions, and they could not simply run to the computer and enter some key words in Google. Suggestions come: They would cut it open to see what's inside (our modern propensity to analyze is alive and well!); they would take out a part and have it chemically analyzed. They would look at some of its material under a microscope. A few participants would not have an inclination to do any of this—they would just let it be. Since this is not the only object of its kind in the world, I can tell them what the results of such analytical procedures would be. After I do this, it at least becomes clear what the object is not. But there is still uncertainty, which leads to repeated sensory examination of the object—perceiving it more closely for possible clues to its nature. Again I answer some questions and they continue to come up with new ideas about what it might be and how they could know whether their conjectures are correct or not.

Over the course of an hour or more the fog begins to clear so that everyone can see the lay of the landscape (or rather, the lay of the mindscape that has developed through all our efforts). And then comes the moment of clarity. Sometimes a participant "gets it" and describes what the thing is; sometimes I fill in the last strokes of the picture and fill out the story of the object. We now know what it is. For everyone the strange thing has found a place in the world that makes sense. This is a moment of relief and satisfaction for most people, although some are disappointed since the tension of not knowing—the sense of riddle—is gone. But I point out that there are still riddles and open questions associated with the object; we have solved one riddle, yet there are others associated with the same object. Any careful study leads us further into the world's "inexhaustible richness" (Merleau-Ponty, 1969, p.139).

The unknown object and our efforts to get to know it make a big impression on people. We have often heard the remark that a course or workshop would have been worth it just for that experience. In response to a survey question, "Is there anything else you would like to share?" one participant, a teacher, wrote (four years after attending a course):

> One of the most remarkable educational experiences I have ever had was when Craig challenged us to explain the nature and identity of the mysterious and unknown [object].... We had no context with which to identify it and thus it provided remarkable and heightened awareness of our powers of observation and reasoning capacities.

A night passes and we return, in memory, to the object. Our focus is now not external observation. We want to attend to our own activity, what we were doing and what was going on in us during the whole process of considering it. So we gather together all the ways in which we were active—what simply seemed to arise in the process and what faculties we were using the whole time. Here is a list that comes together:

- Observing using senses: looking, touching, smelling, feeling weight, listening
- Feeling interest in the thing: a sense of fascination; curiosity; wonder
- Wanting to get to know the object
- Guessing what the object is
- Feeling a riddle
- Picturing, imagining
- Searching inwardly for what it is or what we know that might be like it
- Remembering—we're continually using images and thoughts out of past experience
- Comparing: it's like …

- Categorizing: natural/synthetic, plant/animal, etc.
- Entering a train of thought—logically connecting
- Excluding: we can leave ideas and images behind us (forgetting)
- Focusing and renewing that focus; and also wandering of the mind
- Realizing that thoughts and images just come
- Making associations—sometimes this seems almost random
- Moving back and forth between perceiving and thinking about it
- Noticing that sometimes the search seems ordered; other times it feels chaotic ("imagination overload")
- Feeling that the process is dragging on; getting tired
- Knowing we're not at a conclusion: we always know when we haven't got it yet
- Feeling inner tension
- Feeling excitement and disappointment
- Desiring to keep the process open—not knowing (toward the end of the process)
- Knowing finally the object's place in the world and sensing satisfaction (or not)

So much has occurred, and yet the object seems to have stayed the same. But it is not the same inasmuch as it has become a part of our awareness and we can place it in the world in a way we could not at the outset. The object has become somewhat familiar to us, but not to the degree that we would say, "Oh, it's just a [   ]." Between the first impression of the object and the known object we engaged in a process with the many different features described in the above list. It is quite astounding how much goes on in us.

Normally we don't take the time to reflect on our own process of coming to know. But the unknown object and the time for extended inquiry give us a unique opportunity to see what is involved in this process. It is everything other than simple; it is not neat and orderly. It has times of intense focus and times of letting go. It entails following a train of thought but also using the ability to break free of that train and to be open to new ideas or images that "simply come" along the way. It is a process that is moving, on the whole, in a direction. But it is not a linear pathway to a goal. Rather, it is as if we are coming closer and closer to the object by excluding what doesn't belong to it and by getting clear about everything that has made its existence possible. And for this we need our various capacities of sensing, analyzing, imagining, associating, excluding, remembering, searching mindfully, and so on. We are motivated to carry out these activities—to bring our capacities into play—through our feelings of interest, curiosity, the urge to know, the sense of almost being there, excitement, and wonder. Without such feelings there would be no process of coming to know. And when, for at least a moment, we have reached the conclusion of the process ("I know what it is"), we sense satisfaction. There is a certain quietude and a sense of completion (and for some, disappointment). If we simply stay where we have arrived, we would have stasis. There would be no more movement. But our minds continue to inquire, perhaps now in different directions. Our sensory experiences provide myriad riddles we can explore.

In this exercise we become keenly aware of all that we bring into every experience. Normally we are most aware of the objects of perception and not of our activity that is part of the process. Through such willful exercise our mental life becomes more transparent since our awareness begins to encompass our activity. David Bohm spoke of the task of thought becoming "proprioceptive," by which he means self-aware. Bohm goes so far as to say that the root cause of environmental problems lies in the lack of self-aware thinking:

> The whole ecological problem is due to thought, because we have the thought that the world is there for us to exploit, that it is infinite, and so no matter what we did, the pollution would all get dissolved away.... Thought produces results, but thought says it didn't do it.... Thought is constantly creating problems that way and then trying to solve them. But as it tries to solve them it makes it worse because it doesn't notice that it's creating them, and the more it thinks, the more problems it creates—because it's

not proprioceptive of what it's doing…. I mean that thought is a real process, and that we have got to be able to pay attention to it as we pay attention to processes taking place outside in the material world. (1996, pp. 11, 29, 58)

You could also say: we do so much damage in the world because we don't realize that every perception, every fact, and the result of every action bears our imprint. We are always participants in the world—every appearance (whether a thought, a thing, or the effect of a deed) is something already participated in. Through achieving self-awareness in activity we awaken to our participation.

But this awareness is not enough. The question arises: how can we learn to make our activity increasingly organic and less haphazard? It is the task that Goethe formulates: "If we want to behold nature in a living way, we must follow her example and become as mobile and malleable as nature herself" (2002, p. 56). We have seen how the plant is an organism that turns toward, grows into, and interacts with the world it becomes part of. But the plant also has its own living activity that it unfolds in its interaction with the environment. It develops its life history in a highly ordered and yet flexible way. Our next step is to look at plant developmental dynamics and inquire how it can facilitate dynamic, living thinking.

CHAPTER 3

The Plant as Teacher
of Transformation

Plant Development—The Field Poppy

THE FIELD POPPY (*Papaver rhoeas*) stems from the European Mediterranean area and has spread around the globe where it often grows—unwanted by farmers but enjoyed by the human eye—in agricultural fields. It is an annual plant, meaning it develops each year from seed, and the fully developed plant does not overwinter. It dies, and only the seeds live on into the next season.[1]

Biologist Jochen Bockemühl researched the development of many different plants, including the field poppy, and had specially designed boxes constructed to allow observation of the roots, something that is normally hidden from view (see Appendix). He had the plants drawn or painted at regular intervals. With the help of these illustrations we can gain a concrete picture of the way an annual plant develops from seed to seed.

The seed of the particular plant depicted in Figure 3.1 was sown at the end of April in Switzerland. As we have seen in the previous chapter, the plant takes root quickly, sending down a long primary root that sends side branches into the soil. Three weeks after sowing, a rosette of small foliage leaves has emerged from the very short shoot that extends immediately above the surface of the ground. ("Shoot" is the botanical term for what we more generally call a plant's stem; in a tree we call the main shoot the trunk and the side shoots branches. I will use the terms "shoot" and "stem" interchangeably.)

At five weeks the poppy has grown markedly. The roots have continued to grow and branch further, increasing their contact with the soil environment and anchoring the plant. Above ground the plant has produced more and larger leaves and retains the rosette form in which leaf after leaf grows out of the growing point of the shoot, which, however, remains short, so that the spaces between the leaves (internodes) are small. Viewed from above, the rosette shows that the leaves spiral around the short shoot.

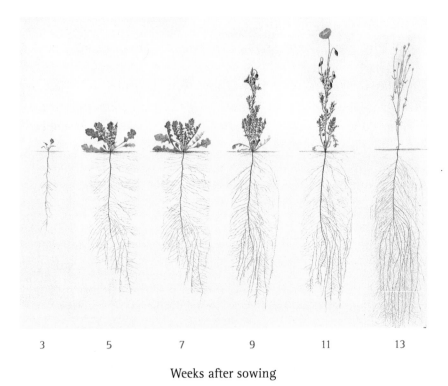

Weeks after sowing

Figure 3.1. The development of the field poppy (*Papaver rhoeas*). The seed from which the plant developed was sown at the end of April in a greenhouse in Switzerland. (After Bockemühl, 1981; reprinted with permission.)

After five weeks the overall form of the root body fills out. The root tips are physiologically active and continue to grow slowly. New leaves develop. In the middle of the rosette, the younger leaves are oriented more upright and they also have a different shape, being more deeply lobed. The lower leaves—the older leaves—spread outward and some lie on the ground. They are turning yellow and will soon decay. Some of the smallest and oldest leaves on the three-week-old plant have already died at this stage.

The above ground part of the plant goes through a large developmental shift between the seventh and ninth weeks. The shoot elongates, growing vertically upward. The leaves still spiral around the stem, but with the elongation the distance between the individual leaves increases. It is now easy to see that the leaves actually do develop one after another. With shoot elongation the basal rosette leaves rapidly die away. The focus of generative growth is moving upward in the plant.

In the nine-week-old plant, flower buds are visible at the end of drooping stalks—a characteristic of poppies. Soon thereafter the first of these flower buds opens. The two green sepals that form the bud casing for each flower fall away and the scarlet red petals unfold to form the red chalice of the blossom. The petals are usually four in number, surround numerous deep purple to black stamens, and the pistil forms the bulbous center of the flower (Figure 3.2).

Figure 3.2. The flower of the field poppy (Papaver rhoeas), with petals unfolded (left) and after petals have fallen off (right). In the center of the flower is the pistil, which is surrounded by the multitude of stamens. (Photos by Dan Tenaglia; missouriplants.com; used with permission.)

The flower is a highly complex structure that appears essentially all at once, in contrast to the gradual unfolding of the foliage leaves. The flower consists of nested circles of highly modified leaves: first the sepals (which are the flower bud leaves), then the petals, then the strongly contracted stamens, and finally the pistil, which forms from fused leaves called carpels. (Why we can consider all these members of the flower to be leaves will be discussed below.)

The petals in each flower are short-lived and usually fall off within a few days. But new flowers open over the course of a few weeks, so that the upper part of the plant presents a changing display of red flower bursts during this time. While the plant flowers, the lower part thins out as more and more of the green leaves turn yellow and die.

At thirteen weeks a virtual skeleton of the plant remains—the main shoot of the plant with its side branches carrying only very few dying leaves and, below ground, the root network. Many of the side branches carry fruit capsules. These capsules develop out of the ovary, which forms the bulk of the poppy's pistil. Pollen from other field poppies is brought by insects to the top of the pistil. A pollen tube develops and grows down to the ovary, which houses a multitude of ovules. When fertilization occurs and the seeds form, the ovary swells into the fruit capsule. Under favorable conditions over one thousand seeds can develop in one capsule. Each seed is a tiny encapsulated plant consisting of two embryonic leaves (cotyledons), an embryonic root, and a tiny cone of embryonic shoot tissue (the growing point) out of which the future shoot, leaves, and flowers will ultimately develop.

So although at thirteen weeks the poppy looks dead, it is in fact bearing an abundance of life. This is not outwardly unfolded life, but rather life encapsulated. Through pollination and subsequent fertilization, the life of one plant has multiplied into many plants in the form of seeds. Each seed carries the potential to develop further as a plant—a potential that in the poppy often goes through a period of dormancy in the soil over the winter. When in the following spring there are appropriate environmental conditions, it will germinate and grow into a seedling. The plant will have come full circle.

In reflecting on the illustrations of the poppy's development— which are of course a surrogate for the study of the actual plant—we can begin to build up a picture of the plant's life cycle as a continuous process. The poppy develops as long as it lives. At any moment in time it is a "whole"—we can recognize it as a poppy—but in another sense it is never whole at any moment. We need to follow the poppy's development over time when we want to see and understand it as a whole organism. It is striking how the plant's structures develop sequentially. The young plant consists of roots, a short shoot, and leaves growing off the shoot to form a rosette. More leaves develop in the rosette and then the shoot elongates, bearing one leaf after the next. In the angle between the base of each leaf and the shoot, a bud, called the lateral or axillary bud, forms. Out of these buds side shoots develop, which can bear leaves and flowers. When the plant begins to flower, many of the foliage leaves have already died. When, in turn, the fruit develops, the petals have fallen off and the stamens wilt. Finally, when the seeds ripen, the rest of the plant—which was the basis for seed development in the first place—is dead.

So when following the poppy's development one sees not only how it grows, gets larger, and finally dies, but also how in growing it is always dying. Growth and death in a plant belong together; they are two aspects of the developmental process: older foliage leaves die as new ones sprout; the upper foliage leaves die as the plant brings forth flower after flower; petals fall off and the ovary swells into the fruit capsule; the fruit and whole plant dry out as the seeds ripen; the dry fruit capsule sets free the seeds. As Tobler wrote in a Goethean spirit, "Life is [nature's] most beautiful invention, and death her craft to make much life" (in Goethe, 1995, p. 4; translation modified by Craig Holdrege).

So the poppy develops continuously during its life span. The continuous stream of life generates new forms and structures over time so that not all are present at any given moment during its life. Growth and decay go hand in hand in this process—when something new develops, something that was essential before drops away.

Leaves: Transformation, Expansion, and Contraction

Just as the plant as a whole goes through a transformation, so do its parts. In herbaceous wildflowers, especially in annuals, there is a marked and yet easily overlooked transformation within the foliage leaves. In botanical terms one speaks of heteroblasty (Dengler & Tsukaya, 2001; Jones, 1999). Figure 3.3 shows the foliage leaves of the field poppy. To the right (b) you see all the leaves of the main stem ordered from bottom to top. At the bottom (c) a complete sequence of foliage leaves from the main stem from a different specimen is shown, but here the leaves are laid out horizontally—from the first rosette leaf (to the left) to the uppermost leaf on the main stem (to the right).

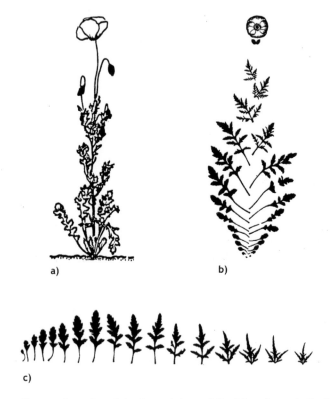

a)   b)

c)

Figure 3.3. The transformation of the size and shape of the foliage leaves in the field poppy (*Papaver rhoeas*). a) The whole plant in full flower. b) All the foliage leaves from the main stem arranged sequentially from bottom to top. All leaves are full grown. The small leaves at the bottom are the first leaves to be formed; the leaves get larger and then reduce in size again near the top of the plant. After the last foliage leaves the two small sepals and the petals, as seen from above, are shown. c) As in b), only a different specimen and leaves arranged horizontally: all the full-grown foliage leaves from the main stem; at the left are the leaves from the bottom of the plant; at the right are the uppermost leaves. (From Bockemühl, 1973, pp. 40–42 (a and c), and Bockemühl, 1981, p. 18 (b); reprinted with permission.)

It is important to note, as hard as it is to believe at first, that all the leaves are full grown. As we already discussed, at any given stage of its development a plant usually does not have all its leaves, so the only way to obtain a full sequence of leaves is to pick leaf after leaf when they begin to wilt and press them.

The small first foliage leaves have a thin, relatively long leaf stalk and a rounded leaf blade that then becomes more oval in shape. Initially, each subsequent leaf is larger than its predecessor, and its form is also different. The leaf margin becomes more strongly toothed and the leaf blade forms lobes. The largest leaf is, in the poppy, in the first third to half of the sequence (this varies from specimen to specimen). When the leaf begins to get smaller again, its form continues to change. The leaf stalk shortens, the lobes emerging from near the base of the leaf stalk become very long and all the lobes increasingly pointed. One could hardly imagine a more radical change in form when one compares the first and last leaves.

When we look at such leaf sequences in courses at The Nature Institute, participants are usually amazed that they had never noticed the variety of leaf types on a single plant, although they may have looked at many weeds and wildflowers. Often the fine articulations of the leaves get overlooked in the overall impression of "green foliage." This is a fine example through which we can become aware of how we tend to overlook what is there to be seen, and how, when we do look carefully, there is a rich diversity in the phenomena that reveals itself.

As Goethe (1995, p. 87) noted, in the formation of its foliage leaves a plant goes through a phase of expansion and then a phase of contraction. It is interesting to realize in this context that the development of each leaf is accompanied by the formation of an axillary bud at its base. When a foliage leaf unfolds — be it small or large, at the beginning of the leaf sequence or at the end—at its base a bud forms that holds the potential to develop into a side branch with leaves and flowers. So each individual foliage leaf is not only part of a larger process of expansion and contraction, but itself shows both expansion, in the formation of the leaf, and contraction or concentration, in the formation of the lateral bud, which might then later expand.

The transformation of the foliage leaves, which ends in the contraction to very small leaves, is followed by an expansion in the unfolding flower. Whereas the change from foliage leaf to foliage leaf shows a relatively gradual transition, the shift from the foliage leaves to the sepals and petals of the flower is usually abrupt and shows little or no transition. In the field poppy, the sepals are still green (meaning they contain chlorophyll and carry out photosynthesis), but they have a distinctly different form and structure than the foliage leaves; they are thick, protective bud scales that possess no stalk. The petals make another quantum leap. They expand outward and present a radically new form of leaf and leaf structure. They are usually delicate, short-lived, and colorful, and do not carry out photosynthesis. In many plants the petals expand far beyond their visual boundary through their scent, which can spread out into the surrounding air and attract insects.

The stamens are the next structure in the flower. They are strongly contracted. When they mature, they release pollen that spreads via wind or insects far into the environment. In this sense the stamens display both contraction in form (morphologically) and expansion through pollen disbursement (functionally).

Out of the pistil develop the fruit and the seeds. The fruit develops through expansion (swelling) of the ovary (and often the surrounding stem tissue), while simultaneously, nested within the ovary, the seeds form through a process of contraction and concentration. As a final expansion, the contracted seeds spread out into the environment via wind, insects, birds, and other animals. When the seeds germinate, a new cycle of expansion and contraction begins. So from a dynamic morphological perspective, the plant develops and transforms through waves of expansion and contraction that are informed by more gradual changes in the foliage leaves and more abrupt changes in the parts of the flower.

Process Thinking

Whereas out of habit we might think of a plant as a "thing," as an object in the world, as a noun, through these observations we begin to see the plant as a process. The plant maintains—or continuously generates—a stream of life of development and transformation. Out of this stream it brings forth its different forms and functions. The plant is continuously changing and developing. The thinking we use to follow and understand the plant as a process cannot be static. It has to move with the processes and transformations.

When we observe the plant's development we are always looking at snapshots of a continuous process. This continuity is not given to us as a sense-perceptible phenomenon. Rather, we gain access to it through our activity of connecting one stage with the next. Because, as William James characterized it, our consciousness is a kind of stream, we can participate in the process nature of the plant through the flow of inner activity that moves and connects the discrete images given to us over time. James wrote about the stream of thought:

> Consciousness, then does not appear to itself chopped up in bits. Such words as 'chain' or 'train' [of thought] do not describe it fitly as it presents itself in the first instance. It is nothing jointed; it flows. A 'river' or a 'stream' are the metaphors by which it is most naturally described. (James, 1950, p. 239)

Within the stream of development in the plant the parts develop in relation to one another. Each foliage leaf in the sequence of leaves, for example, points both to the leaves that came before and to the leaves that are to come. It has a past or a history behind it, and it is developing into the future. Any part we may choose to focus on leads us beyond itself: a part is not understandable out of itself (as a discrete part), but only out of its context within the developmental process. The plant can teach us to look with the eyes of process into the world.

I have a vivid memory of an experience in a college biology class. The professor spent a couple of lectures telling us about how scientists performed a series of feeding experiments with rats. They were trying to find out what substances are necessary and sufficient to sustain animal life. So they tried lots of combinations of substances. Though I don't recall the details he related (and can't find my notes from thirty-five years ago), I'm quite sure he was telling us about the work of E. V. McCollum in the early twentieth century (McCollum & Davis, 1913; Wolf, 1996). McCollum had observed that cattle, depending on the composition of their feed, were healthy or became sick. He wanted to know more about feed composition and its effects on animal health. So he started doing feeding experiments with rats—the ones my professor described. He discovered, for example, that if he fed rats pure protein, milk sugar, a fat source (olive oil or lard), and minerals the animals did not grow after a period of time, and although a female fed this minimalist diet could bear young, it could not produce enough milk to feed them.

McCollum found that if he added butter or an ether extract from egg yolks to the minimum diet, the rats grew again and could reproduce normally. He then examined other substances and discovered that ether extracts from alfalfa leaves, liver, or kidney, for example, had similar positive effects. So what was going on? Evidently the dictum "what you need in the diet is protein, carbohydrates (sugar or starch), fat, and minerals," is not correct. McCollum had discovered that adding these different ether-extracted substance to the diet improved growth. He knew from organic chemistry that ether extraction is used to extract fat-soluble (and not water-soluble) substances out of other substances. He spoke of a "fat-soluble factor A." He could not see, smell or touch this "factor A," but he knew there was "something" that could be derived from a variety of plant and animal tissues that was essential to sustaining life.

This is about as far as my professor went. He then said something like, "Now you know what a vitamin is." McCollum had discovered what was subsequently called vitamin A. I was thrilled.

Why? Because finally a professor had described a process of scientific discovery. Scientists don't just open up food or an organism, peer into it with a microscope and discover minerals, vitamins, enzymes, etc. They have to interact with the organisms. They engage a variety of techniques through which they isolate substances. The way a particular substance is isolated—the interactions with chemicals, temperature, etc.—tells you about its characteristics. To learn what the substance signifies you have to reintegrate it into the life of the organism, and the effects tell you something about the biological nature of the substance. Substances are all about specific processes, and I had gotten an impression of how an important substance like vitamin A arises out of an array of processes, questions, and considerations.

How different this was from memorizing the definition of a vitamin ("any of various fat-soluble or water-soluble organic substances essential in minute amounts for normal growth and activity of the body and obtained naturally from plant and animal foods," http://www.thefreedictionary.com/vitamins) and then moving on. It was memorable for me because it was rare in biology classes to actually get a sense of how a scientific concept takes shape. I learned how scientists work and how scientific ideas grow out of particular research questions that unfold at a particular time within a particular setting. Such teaching shows how a scientific concept (in this case of a particular substance) gradually gains contours within the process of science itself and, as a result, the concept remains open. (After all, scientists learned much more about vitamin A after McCollum's work.) And, importantly, I learned that substances are not "things," but rather part of specific processes.

Later, in my own teaching, I held the memory of this experience as a kind of guide: are you just bringing to students a series of unrelated facts? Are you showing how science is always a process? Are you letting the students experience that every "thing" is embedded in processes and emerges out of them and then back into others? Science *is* process, but how often do we let our students experience that?

Bringing Forth and Letting Go

The plant lives through ongoing growth and decay. At an early stage of its development, for example, the rosette leaves are an absolutely vital part of the plant's existence. Functionally they are carrying out photosynthesis, which enables the roots to grow and branch into the soil. Then, out of the rosette's center the shoot elongates, forming more leaves and finally flowers. The rosette leaves die away. The plant transforms by leaving behind old structures and bringing forth new organs. It does not hold onto the state achieved. Even the root network, which maintains, once established, a fairly consistent form, is continuously changing, since the roots grow at their tips, some roots may die off, and new root branches are formed.

What would it signify to internalize this quality in our thinking? It would mean, for one thing, that we would not hold on to our ideas and concepts as fixed, static entities. Take a definition as an example. A definition is a clearly circumscribed, precise demarcation of a concept's meaning. It may reside in the mind as a finished crystal—beautiful but no longer generative. In this case, the definition "is"; it is the standard against which things are measured—do they fit in the definition or do they contradict it? Used in this way the definition is like an object that excludes other things from it; it does not interpenetrate and meld into something else. A definition epitomizes object thinking. From the perspective of organic process thinking, a definition shows itself as the end result of a specific process—one particular way of understanding that developed out of a particular mode of thought within a particular historical, social or other context. (Owen Barfield's *History in English Words* [1967] gives many vivid examples of how the meaning of words evolve in the course of history.)

A definition is a special case of holding on to an idea or an opinion we have formed. How easily do we move beyond an idea we have learned to cherish? In a course someone remarked once that speaking—formulating a thought—is the formation of something concrete and is like the forming of a leaf. We make a definite statement. When we put out a comment or idea in speaking, that idea can be taken up in conversation. In conversations, misunderstandings and conflicts often arise when we hold on to statements that we or someone else made earlier and we don't realize that the process has gone further. We need to be able to let those particular thoughts go, as hard as that may be. Everyone can sense that such an ability to let go and move on is an essential part of any deeper conversation; it is a form of interaction with the world that is alive.

This living attitude of mind—one that can be learned from studying plant development—allows us to see an idea as an expression of a process. As such, the idea is important and essential, but it has its time, and we would need to be willing, as plant-like thinkers, to let a particular notion fall away as our lives develop further. In this attitude of mind we hold on to our concepts more tentatively because we see our thought life as in process and transformation.

Expansion and Contraction in Learning

The plant's life plays itself out in rhythmical expansion and contraction. In expanding, the plant opens into the environment and interacts with it. In contracting, it forms compact structures such as the seed casing, but also generates germinal points of life, such as the lateral buds, the pollen in the stamens, or the embryo in the seed. Expansion and contraction occur in part over time, as in the development of the foliage leaves and the flower, but also simultaneously as in seed formation (contraction) nested within the maturing, swelling fruit (expansion). The plant lives its life in these polarities.

Do we find something of these qualities in thought? Let me describe a process of inquiry I underwent while studying a particular plant. One early spring in the northeastern United States, I was exploring a wetland and discovered skunk cabbage (*Symplocarpus foetidus*, see Holdrege, 2000b). I was new to the region, and this plant fascinated me. Many hours of observing skunk cabbage over the course of numerous years ensued. I literally went out to the plant again and again. But I also reached out to it through my senses of sight, touch, and smell. I attended to the plant in the different stages of its life cycle, in different specimens, and in different microenvironments. I read everything about skunk cabbage I could get my hands on. I expanded out into the skunk cabbage. This process of going out again and again to get to know something better, studying it from different angles and gaining ever new perspectives on it is like the growth of leaf after leaf in the plant.

But this is not all that is happening. I also have a memory and I develop questions and ideas. I take the expansive experiences in. I remember, ponder, and relate individual experiences to one another. New questions arise that guide my next observation. I also practice picturing the plant to myself, connecting the different growth stages with each other to gain a more vivid image of the way the plant grows. This practice of exact sensorial imagination, as we have seen, is a way of bringing the plant into oneself and enlivening what is perceived. So there is a continuous process of going out, taking in, working within myself, and then going out again. Through the expansive investigation and the internalization and infusion of experience with thought, my understanding of the plant grows. I could also say: the plant of understanding grows in a rhythmical and iterative process of expansion and contraction.

We often speak of the accumulation of knowledge as if it were additive, a process of placing block on block. But we are not simply adding fact to fact or stacking information upon information to build an edifice of knowledge. This is in fact a quite inappropriate analogy—or perhaps we could say it is a mere analogy—because it does not portray the actual process of knowing and learning. What it may fit closest to is what most of us have experienced in school or college: learning unconnected facts by rote memory and placing them side by side (or piling them on each other). The result is often not pretty, and we may only be able to keep the parts cemented together long enough to pass an exam. Then, luckily, the organic process of forgetting—shedding and leaving behind—frees us from the clutter.

When I was learning about the skunk cabbage, each new experience was informed by previous experience, like the foliage leaf that takes on a different form from that of its predecessor, and a new form could develop because the plant had brought forth the previous leaves. Real learning is an organic process since there is coherence and connection between past and present experiences that lead to a more differentiated organism of understanding. As John Dewey stated, "The central problem of an education based upon experience is to select the kind of present experiences that live fruitfully and creatively in subsequent experiences" (1997, p. 28).

Flowering Thought — Insight

Historian and philosopher of science Thomas Kuhn is well-known for his idea of paradigm shifts within science (Kuhn, 1996). He describes vividly such a shift in his own thinking that I like to present in Nature Institute courses (Kuhn, 2002, pp. 15–17). Kuhn was a graduate student in physics preparing a course on science for nonscientists and looking into the development of the science of mechanics, which he wanted to present as a case history. As part of this work he read Aristotle's writings on physics and noted that "Aristotle appeared not only ignorant of mechanics, but a dreadfully bad physical scientist as well. About motion, in particular, his writings seemed to me full of egregious errors, both of logic and of observation."

Kuhn did not remain with this harsh conclusion; he had the sense that he wasn't understanding Aristotle correctly. He knew Aristotle was the founder of modern logic and had laid the groundwork for biological sciences. "How could his characteristic talents have deserted him so systematically when he turned to the study of motion and mechanics?" And "why had his writings in physics been taken so seriously for so many centuries after his death?" Kuhn was unsettled: "Those questions troubled me."

He continued to "puzzle over" Aristotle and reread what he had studied before. And then something happened:

> I was sitting at my desk with the text of Aristotle's *Physics* open in front of me and with a four-colored pencil in my hand. Looking up, I gazed abstractedly out the window of my room—the visual

image is one I still retain. Suddenly the fragments in my head sorted themselves out in a new way, fell into place together. My jaw dropped, for all at once Aristotle seemed a very good physicist indeed, but of a sort I'd never dreamed possible. Now I could understand why he had said what he'd said, and what his authority had been. (Kuhn, 2002, p. 16)

In order for this shift in understanding to occur, Kuhn had to drop the Newtonian lens through which he had been considering Aristotle. It meant opening himself to the idea that "perhaps [Aristotle's] words had not always meant to him and his contemporaries what they mean to me and mine." In his sudden realization, Kuhn was able to get inside Aristotle's take on things. He realized that Aristotle did in fact mean something very different from Kuhn and his Newtonian mind when he spoke, for example, of motion or matter. It became clear that while "in Newtonian physics a body is constituted of particles of matter, and its qualities are a consequence of the way those particles are arranged, move, and interact," for Aristotle qualities are primary. There is, for Aristotle, change in the world not because matter changes but because qualities (such as heat, moisture, and color) mold matter, which in itself is essentially devoid of qualities. This is a wholly different way of viewing the world, and only when Kuhn was able to get inside this worldview and leave behind his own did he understand Aristotle. What had previously been for Kuhn a multitude of strange and in part erroneous views "fell into place together" and revealed itself as a coherent whole.

Kuhn's whole way of viewing Aristotle shifted, not as a gradual process but in one moment. He goes on to say that this kind of experience is exemplary of revolutionary changes that happen in human understanding as the beginning of a paradigm shift: "It involves some relatively sudden and unstructured transformation in which some part of the flux of experience sorts itself out differently and displays patterns that were not visible before" (Kuhn, 2002, p. 17).

I would call this an example of the blossoming of a new idea. It builds on a process of study: Kuhn engages in Aristotle. He comes to certain strong views (leaves of thought), but he doesn't hold onto them. His other knowledge of Aristotle and his intention to do Aristotle justice prove stronger, so he continues to study *Physics*. Then all of a sudden—and so striking that the exact external conditions of the insight remain clearly in memory—the new configuration of knowledge appears. He has grasped as a new whole what before were only parts of the picture. At once the coherent new form—the blossom— appears. Generally, we can say that all "aha" experiences, in which suddenly something becomes clear to us, are such flowers in our mental life.

David Bohm (2005a) characterizes the moment of insight:

> Suddenly, in a flash of understanding, involving in essence no time at all, a new totality appears in the mind.... This new totality is at first only *implicit* (i.e. unfolding) through some mental image which, as it were, contains the main features of the new perception spread out before our "mental vision." Perception involving this display, which is inseparable from the act of primary perception itself, is what may be called *imaginative insight* or creative imagination. (p. 54; his emphasis)

In a plant the flower is developing during the leafing phase, but its actual appearing always brings a kind of surprise; just through the study of the foliage leaves of a plant one cannot know what kind of flower will develop. It emerges out of the rest of the plant, without which it would never come about, but in a sense you cannot know beforehand what is coming. This is also true of moments of insight. They grow out of past work, and often the most profound experiences arise when one has worked long and hard on a particular problem or wrestled with certain questions. And then, as if out of nowhere, the insight comes—perhaps when you are taking a walk and least expect it.

While the flower arises as part of the organic metamorphosis of a plant, it does not "automatically" arise, just as a moment of insight does not necessarily follow upon a pathway of study. The plant needs certain kinds of conditions—soil, light rhythm and intensity, etc.—

that allow it to develop. Are there conditions that can foster such moments of insight? As Douglas Sloan suggests, "All genuinely new knowledge comes by means of passionate, energy-filled insight that penetrates and pierces through our ordinary ways of thinking. The function of insight is twofold: to remove blocks in our customary and fixed conceptions of things, and to gain new perceptions" (1983 p. 141).

To remove the blocks of habits of thought, we can take a lesson from the way a wildflower grows. For example, we can create times when we draw back (contraction in the plant) from a train of thought, since by keeping a continuous focus we often get cramped and confined by the pathway we have taken. If we pull back and shift into a mode of open awareness of what may come, we heighten our receptivity. This is akin to the practice I described in the previous chapter as "sauntering of the eyes." In this context the emphasis is on the letting go and waiting to see what may come—we create the inner space for new perceptions that can appear to the mind's eye. David Bohm describes how the "creative state of mind" is "like that of a young child … always open to learning what is new, to perceiving new differences and new similarities, leading to new orders and structures, rather than always tending to impose familiar orders and structures in the field of what is seen" (2005a, p. 21).

Insight cannot be forced, as we know all too well. Interestingly, moments of insight often come after we've "slept on something." Sleep is a time of forgetting and freeing ourselves from all the phenomena and thoughts of the day. It allows us to awaken to each day afresh.

Along the way I will bring further examples of flowering insight.

Leaf Metamorphosis

Figure 3.4 shows the foliage leaves of six different species. We see a great diversity in the forms and in the course of leaf development up the stem. Wild radish (*Raphanus raphanistrum*) forms large rounded leaves that are only slightly lobed, while common ragweed (*Ambrosia artemisiifolia*) has finely dissected leaves. Its two cotyledons (the first pair in the sequence) have round blades but already the first foliage leaf is deeply lobed; only through the rounded tips of the lobes and in comparison to the leaves that follow do we recognize it as an early leaf. Or we could also say: seeing a strongly lobed first foliage leaf foreshadows things to come—that this plant will have strongly differentiated leaves.

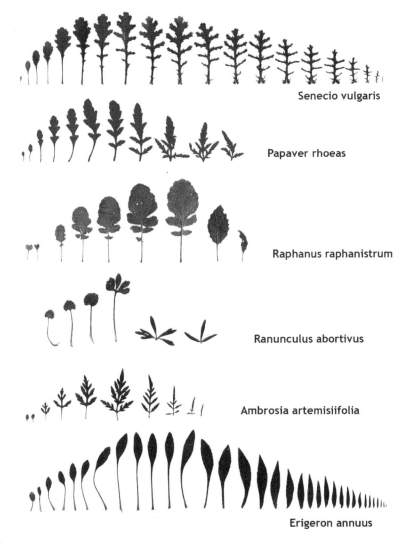

Figure 3.4. Leaf sequences. Each sequence represents the fully grown foliage leaves on the main stem of one plant; on the left of each row are the first leaves and on the right the last leaves before the flowers. (Scans of pressed leaves; Craig Holdrege.)

In contrast, the leaves of daisy fleabane (*Erigeron annuus*) never form lobes at all. The blades of the first foliage leaves are oval-shaped and pointed at the tip. The leaves then develop long stalks and narrow blades, and the stalk gradually disappears as the leaves become lance-shaped and ever smaller. The most surprising transformation is in the small-flowered crowfoot (*Ranunculus abortivus*). Whereas in the other species there is a more or less gradual transformation, in the crow-foot there is an abrupt shift. It is as if leaves were missing between the fourth and fifth leaf. But this is not the case; it is typical for this species to have a form gap between the basal leaves and the first leaf on the up-right stem. There is also no gradual contraction; it occurs in one step.

Through this comparison we are led to a dynamic concept of the foliage leaf. *Merriam-Webster's Collegiate Dictionary* definition of the leaf is interesting in this context: "a lateral outgrowth from a plant stem that is typically a flattened expanded variably shaped greenish organ, constitutes a unit of the foliage, and functions primarily in food manufacture by photosynthesis" (p. 662, 1999, 10th edition). This is an encompassing definition and points at essential aspects of the leaf. But what it does not capture at all is the dynamic quality of the leaf. To behold this quality one has to follow developmental sequences, attend to the concrete variety, move with the mind between the different forms, and see them all as expressions of leaf-forming potential. We begin to see how leaves can be very different from each other and yet all still be leaves. And there is dynamic order as well, since each plant goes through a leaf transformation with expansion and contraction, although each does it in its own unique way.

Metamorphosis in the Flower

We have seen how the plant moves into another realm when it flowers. Petal, stamen, and pistil (which is made up of one or more carpels) differ remarkably in form and function from the foliage leaf. Yet, like the foliage leaves they are outgrowths of the stem (shoot). The continuous life of the plant produces, sequentially, its different parts. It is the one plant, being itself, but differently in the variety of its parts.

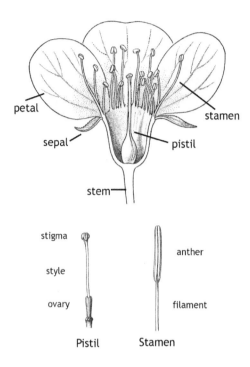

Figure 3.5. A schematic representation of the parts of a flower. (After Troll, 1957, pp. 15, 54, and 62.)

When Goethe studied plants, instead of only naming their parts and identifying their structural relations (Figure 3.5), he saw the parts as expressions of what he called the same "organ":

> The organ which expanded on the stem as leaf, assuming a variety of forms, is the same organ which now contracts in the sepals, expands again in the petal, contracts in the reproductive apparatus, only to expand finally as the fruit.... Thus we can say that a stamen is a contracted petal or, with equal justification, that a petal is a stamen in a state of expansion; that a sepal is a contracted stem leaf. (Goethe, 1995, pp. 96–97; translation modified by Craig Holdrege)

He came to this insight not only through knowing that all these parts are outgrowths of the stem, but through the study of the parts themselves, natural variations of the parts within one species, and through comparison of different species. The result was his *Metamorphosis of Plants*, written in 1790.

Figure 3.6. From foliage leaf to stamen in the peony (*Paeonia* sp.). The cultivated peony shows like almost no other plant the transition from one leaf type to another.
1: upper foliage leaf, beginning to broaden at its base into a sepal
2: an intermediate form that is mainly sepal with remnant of an upper foliage leaf
3: a sepal; the pink petal color is already apparent in the outer margin of the sepal
4: a fully developed petal
5: irregularly formed petals
6: intermediate forms that are part petal and part stamen
7: a fully developed stamen
(From Suchantke, 2009, p. 63; reprinted with permission.)

It is not difficult to recognize that the sepals, which are the bud leaves that enclose the flower before it opens, are highly contracted leaves. They are usually (!) green, carry out photosynthesis, and often resemble to some degree the small uppermost foliage leaves. Although in color, texture, and scent, petals usually show little resemblance to the foliage leaves, they do have the character of leaves.

The resemblance to something leaf-like virtually disappears in the stamens. But some plants show transitional forms even here, such as the many-petaled garden form of the peony (Figure 3.6). In the peony, the upper foliage leaves contract (1), and their base widens. In this way they gradually become sepals (2). The outer edges of the sepals take on a red color, hinting at what is to come in the petals. The petals (4) are pink in color and have a delicate texture that is so different from that of the robust sepals. Toward the middle of the flower the petals become smaller, vary in shape (5) and some develop half as petal and half as stamen (6) before you discover the typical contracted stamens (7).

Not many plants show such a step-wise metamorphosis within the flower. Another example of a wild plant that has, as a rule, transitional forms between petals and stamens is the water lily (*Nymphaea* sp.; see Figure 3.7). The petals become ever narrower toward the center of the flower, and at the tip small anthers appear that increase in size as the petal shrinks to a filament; the result is a typical stamen.

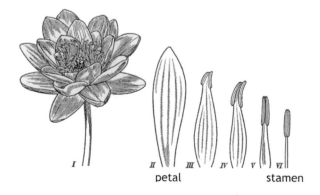

Figure 3.7. Transitional forms between petal and stamen in flower of the white water lily (Nymphaea alba). (From Troll, 1957, p. 17)

The pistil forms the center of the flower and inside the ovary the seeds develop. When the pistil is simple, it consists of one so-called carpel. This is the case in peas (Figure 3.8). When pollination has taken place, the ovary begins to swell and turn into the pea pod. This pod is, botanically speaking one carpel—a leaf-like structure with a midrib and two surfaces that have folded and grown together. A seam forms where the two carpel halves join. On the inside of this seam, which corresponds to the margin of a leaf, the ovules arise and develop into seeds. So it is at the margin of this self-enclosing "fruit leaf" that the seeds for the next generation develop! (In more complex flowers the pistil consists of multiple carpels that can be separate, partially fused, or completely fused. For example, each of the many sections of an orange that we pull apart after peeling the fruit are individual carpels.)

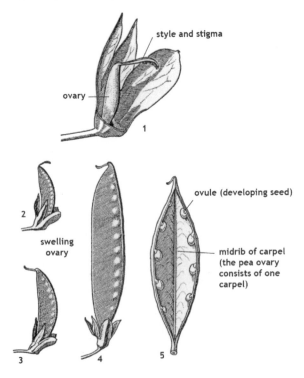

Figure 3.8. Fruit (pod) development in the pea (*Pisum sativum*). The pistil consists of one carpel; the ovary portion of the carpel swells and inside this "fruit leaf" the seeds form. (Altered, after Troll, 1957, p. 49.)

Occasionally one finds transitional forms between petals and carpels, and more rarely between stamens and carpels (Figures 3.9 and 3.10). Stamens and carpels are polar structures in the flower and the most one-sided in their development; when they reunite through pollination and fertilization viable seeds form—the next generation.*

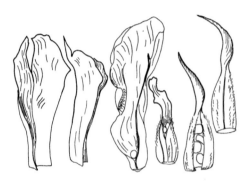

Figure 3.9. Intermediate forms between petal and carpel ("fruit leaf") in the peony (*Paeonia* sp.). At the left is a petal with a somewhat thickened base. The next forms are part petal and part carpel. The two forms at the right show the carpel with ovules, but the upper tip, is colored like a petal. (From Bockemühl, 1982, p. 121; reprinted with permission.)

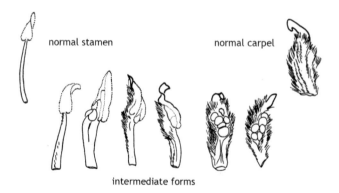

normal stamen   normal carpel

intermediate forms

Figure 3.10. Intermediate forms between stamen and carpel in the peony (*Paeonia* sp.). (After Bockemühl, 1982, p. 118; reprinted with permission.)

---

* In studying life you get used to the fact that there are no fixed rules; there *are* plants—such as the common dandelion—that can form viable seeds without any fertilization at all! This is called apomixis.

A Shape-Shifting Proteus

All these phenomena show the inner connection between the different parts of the plant, or I could also say, they show unity in the diversity of structures. This is a dynamic unity. It is the same life of the plant that expresses itself as cotyledon, foliage leaf, sepal, petal, stamen, and carpel. Goethe called what shows itself differently in the different parts of the plant "leaf":

> While walking in the Public Gardens of Palermo, it came to me that in the organ of the plant which we are accustomed to call the *leaf* the true Proteus lies hidden, who can conceal or reveal himself in all formations. Forward and backward, the plant is always only leaf. (Goethe, 1982, p. 363; translation modified by Craig Holdrege)

Why "true Proteus"? Proteus is a Greek god of the oceans; he is, as Homer describes in *The Odyssey*, "the ever truthful Old Man of the Sea" (1967, Book 4, lines 345 ff.). Knowing the depths, he can arise and take "the form of all creatures that come forth and move on the earth, he will be water and magical fire" (Book 4, lines 417–418). Proteus, like the life of the plant that Goethe calls "leaf," is a shape shifter! In this way he can deceive, and yet, paradoxically, he is always called the "ever truthful." But he deceives only inasmuch as a person cannot recognize him in his different disguises. Like the different parts of the plant, he manifests in ever-changing appearances and yet, like the plant, always remains himself. What a wonderful image Goethe chose to characterize the ever-changing, yet ever-truthful unity of the plant.

There has been a good deal of discussion within academic botany about the wisdom of Goethe's choice of the term "leaf" to name the unity of the plant and about how one might better characterize plant metamorphosis (see, for example, Kaplan, 2001; Kirchoff, 2001; Sattler & Rutishauser, 1997; Schilperoord, 2011). In the context of this study, a definitive conclusion is much less important than the fact that the discussions show that the plant itself is so dynamic and reveals so many variations (leaves that are shoot-like, leaf-like shoots, etc.) that it continually challenges researchers to drop hard and fast categories and to think in dynamic terms.

Goethe's idea that the plant is "forward and backward leaf" was growing in him throughout his 1786–1787 journey through Italy. Then, during his visit to a public garden in Palermo, Sicily, the idea came to him—flowered for him—in all clarity. He had discovered the leaf to be the true Proteus in the plant. Looking back on this experience he writes:

> Only a person who has himself experienced the impact of a fertile idea ... will understand what passionate activity is stirred in our minds, what enthusiasm we feel, when we glimpse in advance and in its totality something which is later to emerge in greater and greater detail in the manner suggested by its early development. Thus the reader must surely agree that, having been captured and driven by such an idea, I was bound to be occupied with it, if not exclusively, nevertheless during the rest of my life. (Goethe, 1989, p.162; written in 1831)

Goethe glimpses the idea of the leaf nature of the plant "in its totality"—the flower of insight. While the flower itself—the concrete experience in Palermo—falls away, the experience bears fruit throughout his long life. We often say an idea bears fruit, and little do we realize how apt a metaphor that is. It's not the idea as such that lives on, but what grows out of it. Or, one could say, the idea lives on in what grows out of it. And one idea can bear many seeds that, in the appropriate environs, can generate ever-new life.

Unity in Diversity—Living Ideas and Wholeness

The dynamic Goethean view of the plant is easily misunderstood, which is an expression of the difficulty the modern intellectual mind has in grasping dynamic ideas (Gädeke, 2000; Harlen, 2005; Kranich,

2007). As the philosopher and historian of science Ernst Cassirer says about Goethe, "There prevails in his writings a relationship of the particular to the universal such as can hardly be found elsewhere in the history of philosophy or of natural science" (cited in Bortoft, 1996, p. 78). Philosopher Henri Bortoft (1996) has gone to great pains to articulate Goethe's approach and to clarify misunderstandings (see also Brady, 1998).

At the one pole of misunderstanding lies the notion that Goethe abstracted from the different phenomena and formed the abstract general category "leaf," under which all instances can be subsumed. Such an idea is a "unity from which all differences have been removed" (Bortoft, 2007). At the other pole of misunderstanding is the idea that Goethe's "leaf" is in reality a physical organ out of which the different kinds of leaves arise. For an object thinker, "leaf" must be something clearly circumscribed and tangible. The modern version of this misconception is found in molecular biology when geneticists equate Goethe's idea with genes that affect leaf development (Goto, Kyozuki, & Bowman, 2001).

But as Bortoft writes, Goethe's conception "is neither internally subjective (a mental abstraction) nor externally objective (a primitive organ)" (Bortoft, 1996, p. 79). Goethe strove to see the general within the particular, never separating them into two worlds; this was the singular quality of his mind. "Instead of a movement of mental abstraction from the particular to the general, there is a perception of the universal shining in the particular" (Bortoft, 1996, p. 79). It is this way of apprehending—where the general lights up within each concrete instance—that allows us to form dynamic, living concepts. Goethe describes the specific quality of his dynamic view as follows:

> If we look at all these *Gestalten* [forms], especially the organic ones, we will discover that nothing in them is permanent, nothing at rest or defined—everything is in a flux of continual motion. This is why German frequently and fittingly makes use of the word *Bildung* [formation] to describe the end product and what is in

process of production as well....* When something has acquired a form it metamorphoses immediately into a new one. If we wish to arrive at some living perception of nature we ourselves must remain as quick and flexible as nature and follow the example she gives. (1995, pp. 63–64)

Here we see how Goethe's elastic mind was stimulated by the dynamic nature of nature to become more attuned to dynamism. Rudolf Steiner, the first editor of Goethe's scientific work and an astute observer of Goethe's methodology, characterizes Goethe's thinking in the following way:

Goethe's thinking was mobile. It followed the whole growth process of the plant and followed how one plant form is a modification of the other. Goethe's thinking was not rigid with inflexible contours; it was a thinking in which the concepts continually metamorphose. Thereby his concepts became, if I may put it this way, intimately adapted to the process that plant nature itself goes through. (1987, p. 30; translation by Craig Holdrege)

You could say that Goethe's thinking was naturally alive and adaptable, and therefore wellsuited to understanding life, but he did not just utilize his gifts, he developed and enhanced them through the methodical study of nature. Goethe expresses succinctly the involved process of forming dynamic concepts (this essay was originally written in 1795):

If I look at the created object, inquire into its creation, and follow this process back as far as I can, I will find a series of steps. Since

---

* It is interesting that the world "Bildung" in German also means education. What could be more formative, in an educational sense, than learning from dynamic and formative processes in nature?—CH

these are not actually seen together before me, I must visualize them in my memory so that they form a certain ideal whole. At first I will tend to think in terms of steps, but nature leaves no gaps, and thus, in the end, I will have to see this progression of uninterrupted activity as a whole. I can do so by dissolving the particular without destroying the impression…. If we imagine the outcome of these attempts, we will see that empirical observation finally ceases, inner beholding of what develops begins, and the idea can be brought to expression. (1995, p. 75; translation modified by Craig Holdrege)

This way of thinking leads us to a perception of wholeness and connectedness in nature. It is, in essence, an ecological approach. To think ecologically is to think in terms of connections and dynamic relationships. But ecologist Frank Golley points out that although ecology is the science of the study of connections and should deal with wholes, it has the problem of

reinforcing mechanical and abstract metaphors. How does one speak about connection in a culture of separation and isolation? I don't know. I do not feel that I have completely solved the problem…. We shift our attention from one part to another, but we cannot easily sense the whole. (1998, pp. 231–32)

The Goethean approach provides a way out of this dilemma, because it helps us to see both the uniqueness of life forms as well as dynamic patterns within that uniqueness. Each concrete leaf points beyond itself and shows itself as connected with the other manifold expressions we call leaves or petals or stamens. Through this kind of thinking, exercised through phenomena-based plant study, we can hone our capacity to see dynamism, connectedness, and wholeness in the world.

Some Implications

What can we say so far, in the way of a summary, about the implications of this way of studying plants?

First, it leads us to an experience of concrete phenomena and their variations. We learn to see and appreciate through concrete instances the immense diversity of the living world. Our sense of wonder and our feeling of connectedness to the world grows. As one of the course participants wrote in her course evaluation, "I believe this may be the greatest gift a course can offer as it stimulates our interest and wonder in the world and spurs us to delve deeper into its riddles."

Second, we have come to see how the development of a wildflower and, in particular, leaf transformation exemplifies the dynamic nature of life. In his book on ecological literacy Frank Golley makes the point that dynamism is a central feature of life on earth and it "requires that we give up the notion that the world and life are static and fixed. No objects in nature are unchanging" (1998, p. 230). The plant is a teacher of dynamism, prodding us to look beyond fixity to transformation.

Third, we learn to see how the process of inquiry or investigation itself can be as dynamic as the plant's transformation: forming and reforming ideas in a process of investigation; letting notions fall away as inquiry proceeds; working in a rhythmical interplay between exploring and reflecting, so that a plant of thought or understanding begins to grow; having moments of insight in which one sees new connections and relations—the flower of thought; these insights transforming into seeds for further growth that can unfold in new contexts. The growth of knowledge itself has deeply plant-like characteristics.

Fourth, we learn to see wholeness, that is, how in nature every part is part of some larger dynamic whole. Plants can help us discover unity as a dynamic quality revealing itself in the diversity of phenomena. We learn to see connections between things that at first may appear to be separate from each other. We begin cultivating a truly ecological view of the world. And this understanding grows more differentiated when we look closely at the relation of the plant to its environment.

CHAPTER 4

The Plant as Teacher of Context

ONE SUMMER I WAS OBSERVING plants growing in edge environments—the transitions between wooded areas or hedgerows and pastures—and I made an observation that struck me. I was looking at plants that bordered a pasture, and then down along a roadside, looked at plants growing in the roadside ditch. I saw some of the same species, but they looked markedly different (Figure 4.1). The plants from the pasture edge were much smaller in height, thickness of stem, and leaf size. In contrast, the ditch plants were robust and showed effusive growth not only in their large size but in their tendency to form many side branches. The two microenvironments were only about fifty meters apart.

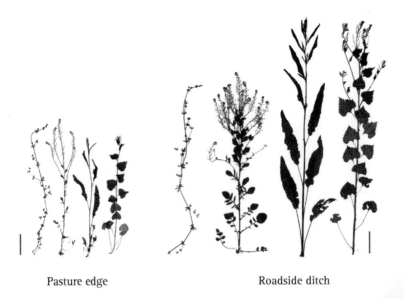

Pasture edge   Roadside ditch

Figure 4.1. Specimens of four different species in two different microenvironments. From left to right in each group: wild madder (*Galium mollugo*), winter cress (*Barbarea vulgaris*), curled dock (*Rumex crispus*), and garlic mustard (*Alliaria officinalis*); scale bars: 10 cm. (Silhouettes of pressed plants; Craig Holdrege)

I was seeing, in a sense, the environment through the lens of the size and form of these plants. Each species was clearly recognizable as such in both microenvironments. But the plants in each microenvironment had a specific "signature"—that of the environment they were growing in. What was it in the microenvironments that made such a difference for the plants?

The ditch is a moister environment, receiving runoff from the road and from the fields that border the ditch. The runoff carries a good deal of dead organic matter with it that often gets caught by fallen twigs and branches and then decomposes. The soil in the ditch where these plants were growing was mucky and loose, a stark contrast to the compact clay-rich soil of the pasture edge. The ditch plants received more direct illumination during the day, having a southerly exposure; the pasture-edge plants had a southeastern exposure. Those are some of the differences between the microenvironments that I detected. I could have carried out a much more detailed analysis of both the differences between the microenvironments (for example, soil pH, salt content since the road is salted in the winter, etc.) and between the plants (overall biomass, root structure, etc.), which would have enriched the picture. But such an analysis would not "explain" the differences; it would simply contribute to a richer and more differentiated picture of the situation.

In The Nature Institute courses we often consider such examples that we find on short excursions into the surrounding area. Most people realize, for the first time, that plants of the same species actually grow in markedly different ways depending on the place one finds them. And then there is the remarkable observation that, as the above example with the four species shows, plants of different species respond to differing environmental conditions in a similar way. I want to look more carefully at this capacity of the plant.

A Plant in Its Context

Figure 4.2 shows two specimens of wall lettuce (*Lactuca muralis*; Bockemühl, 1985). What might the differences be related to? Age or environment (water, soil, light, etc.), or are they different genetic varieties of this species? The urge to know what "causes" the differences is strong. But I want us to withstand that urge and first take time to observe the plant. Why do we always want to explain something before we have considered it with care? In a certain sense we run the danger of manhandling phenomena by getting an answer to a question without having entered into any authentic conversation with the phenomena first. So, we can hold back and begin to describe what we see.

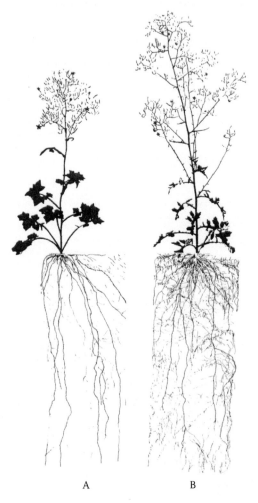

A   B

Figure 4.2. Two specimens of wall lettuce (*Lactuca muralis*). Both plants are flowering. See text for further description. (From Bockemühl, 1985, p. 111; reprinted with permission.)

The smaller plant on the left (A) has a number of long roots with short side-roots, whereas plant B on the right has many more roots due to intensive branching. The foliage leaves of plant A are long-stalked and have a relatively large blade. In contrast, the leaves of plant B are more strongly divided. We could say that plant A forms foliage leaves typical of the early stages of leaf development (see previous chapter), in which the leaf stalk elongates and the leaf blade spreads out to form a more uniform surface, while the leaves of plant B represent more the middle and late phase of leaf development, in which the leaves further differentiate and then become lancet-like. Nonetheless, both plants went through a complete transformation of leaf form in their development. The leaves also have a different orientation on the two plants: in plant A the leaf stalk grows up and out from the main stem at an angle between 30 and 60 degrees, while the leaves in plant B are nearly perpendicular to the stem and curve downward.

In respect to the overall proportions of the aboveground plants, the flowering portion of plant A is relatively compact and small in comparison to its leafing part. Moreover, the lower, leafing part of the plant and the upper, flowering part of the plant are distinct from each other. In contrast, the flowering portion of plant B is expansive. It forms twice as many flowering branches coming off the main stem and branching begins much lower down on the plant. It has nearly twice as many flowers as plant A.

From a quantitative perspective we can see that plant B forms more roots, more leaves, more side branches, and more flowers. Qualitatively, we can say that it shows greater differentiation and branching. The leaves are strongly divided, while the stem and roots show rich branching. Plant A, in contrast, is simpler and more contained. The leaves spread into larger surfaces and do not divide into multiple leaflets. Similarly, the flowering side branches form a tight sphere and the roots branch little, extending out much less into the environment than in plant B.

What is remarkable here is that in both plants—each in its own way—roots, leaves, and stems grow in a unified similar fashion. The unity of the plant shines through all its parts. In plant B all parts show vigorous growth and greater differentiation and branching; in plant A there is more contained growth and the development of simpler forms. This demonstrates how the plant is a unity that members into different parts, each with its own characteristics but nonetheless revealing the whole to which it belongs.

I usually tell participants in a course or workshop at this juncture that we are dealing with an environmental difference—this was an experiment in which the scientist wanted to see how different specimens of the same species respond to different conditions. What could such conditions be? Poor soil or rich soil; ample water or little water; shade or sunlight; high pH or low pH; different planting times. Such are the ideas that people suggest. Sometimes I open up the inquiry to speculation: "Well, what do you think the difference might be and why do you think that?" Many different suggestions follow.

At some point I reveal that one of the plants grew in half-shade and the other in full sunlight. But which is which? What follows is highly interesting. You get arguments for both cases. There are good reasons to think that plant A is the shade plant and good reasons to believe that it grew in full sunlight. And the same with plant B. For example, viewed as a shade plant, plant B looks as if it might be reaching for the light and its roots need to spread out and branch to find nutrients to compensate for the lack of sunlight. No! comes a response, it's exactly the opposite: Plant B is growing in full sunlight; it has plenty of light so it can spread out through branching and can form more roots. Plant A, in this view, would be the shade plant that doesn't grow so large and branches less because it has less light to do so, and its leaves spread out surface-like to catch more light. But others counter that plant A looks more harmonious and well-proportioned, an expression of more light, while plant B looks more straggly as if in need of more light.

So both perspectives have their merits. This is a fruitful exercise in speculation: it shows how easily we can come up with explanations that are in and of themselves coherent. Coherent explanations may contradict each other and may have little to do with reality. Voting on which plant is which may be amusing, but the result would have, thank goodness, little bearing on the issue at hand. The only way to solve the dilemma is through the phenomena themselves. We have to, in this case, go back to the experiment and inquire which plant grew in which conditions. The answer is: plant A on the left grew in half-shade while plant B on the right grew in full sunlight.

We leave the field of speculation and explanation and move to pondering what would be an adequate, plant-sensitive way of formulating the relation of the plant to its environment. We see the plant as an organism that has the ability to integrate itself into its environment in such a way that it reflects in all parts its relation to that particular environment. It takes in from its environment and forms itself through this interaction, and the resulting forms reflect the environment. The shade plant reveals its interaction with its shady environment just as the plant growing in full sunlight shows us in its leaves, roots, and branching how it has interacted with its light-filled environment. What is striking is that those parts of the plant not exposed to light—the roots—show nonetheless the relation to the shady environment. The plant responds as or interacts as a whole with its environment, and each part of the plant expresses in the manner of that part the qualities of the environment it is growing in. Light calls forth greater differentiation, while in shadier conditions the plant stays simpler and ramifies less strongly into the environment. In this way, the shade plant shows us what shade means to wall lettuce just as the sun plant shows us what full sunlight means to wall lettuce. Each specimen of wall lettuce shows us at the same time both its qualities as a species and the qualities of the environment it has grown in.

Because this way of understanding and expressing the relation of a plant to its environment is unusual, I will present some more examples of plant growth in relation to light and shade. By immersing ourselves in a variety of phenomena we can both broaden and deepen our understanding of the plant-environment relation.

Light and Plant Form

Botanist Cynthia Jones (1995) did experiments with gushaw (*Cucurbita argyrosperma* subsp. *sororia*), a wild relative of the pumpkin that grows in western Mexico. She grew plants in partial shade and full sunlight in experimental field plots in California. Figure 4.3 shows leaves from sun plants and from shade plants. As in the wall lettuce experiment, the shade leaves tend to be larger and form less deeply cut lobes, while the sun leaves remain smaller and form more lobes and more deeply cut lobes. The sun plants formed side branches while the shade plants formed none or only a few. Although gushaw typically grows in full sun and wall lettuce grows mainly in edge environments in half-shade, both plants respond in similar ways to changing light conditions.

Jones also investigated the development of the leaves from their origin as undifferentiated primordia to fully developed leaves. She found that leaves at comparable positions on the sun and shade plants were virtually indistinguishable at the earliest stages of leaf development until they reached the length of about 1 mm, when differences became apparent. This led her to conclude that the basic pattern of

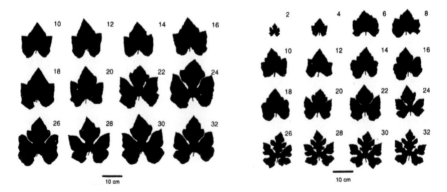

Figure 4.3. Shade leaves (left group) and sun leaves (right group) from gushaw (*Cucurbita argyrosperma* subsp. *sororia*). The numbers indicate the nodes from which the leaves grew, beginning with the node nearest the root. The leaves with the lower numbers come from the lower part of the plant and are the oldest leaves; the leaves with the higher numbers come from higher up and are younger; all leaves are full grown. (Silhouettes of leaf blades, from Jones, 1995, p. 351; reprinted with permission.)

leaf transformation is maintained as "an intrinsic developmental property of the shoot" regardless of the growing conditions, and then the "plastic responses of individual leaves" set in (Jones, 1995, p. 357).

In trees there is often a substantial difference between the sun leaves and the shade leaves on one and the same tree. Figure 4.4 shows the sun and shade leaves of a red oak tree (Quercus rubra) that grew in a forest in upstate New York. Again we see the difference between the spreading, large shade leaves and the smaller, deeply lobed sun leaves.

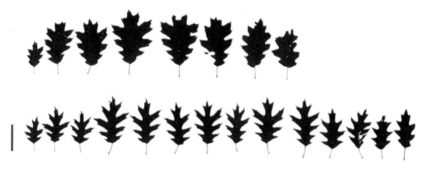

Figure 4.4. Shade leaves (top row) and sun leaves (bottom row) from the same red oak (*Quercus rubra*). The shade leaves are from one terminal twig of a side branch about 16 m (49 ft.) high on a 24 m (73 ft.) high tree. The sun leaves are from a terminal twig at the top of the tree. Both sequences show all the leaves growing on the respective twigs. The leaf at the left is from the base of the respective twig while the leaf on the right is from the tip of the twig. Scale bar: 10 cm. (Silhouettes of pressed leaves; Craig Holdrege.)

Differences between sun and shade leaves in one tree go far beyond size and shape. A research team found, for example, in the European beech (*Fagus sylvatica*) that in addition to being smaller in surface area the sun leaves are thicker, have lower water content, a higher density of pores (stomata) on the leaf undersides, and contain more chlorophyll per leaf unit area (Lichtenthaler, et al., 1981). However, since a typical shade leaf is larger, it possesses overall more chlorophyll than a sun leaf. Microscopically, the leaf structure differs markedly. The sun leaves form a thicker layer of so-called palisade cells, the main cells that carry out photosynthesis (see Figure 4.5). Physiologically, the sun leaves use more carbon dioxide per unit time, pointing to a higher rate of photosynthesis. (For additional studies, see Ashton & Berlyn, 1994; Eschrich, et al., 1989; Hanson, 1917; Niinemets, et al., 1999; Wylie, 1951.)

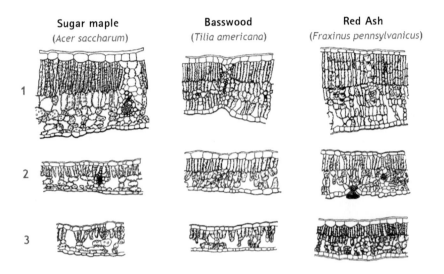

Figure 4.5. Microscopic cross sections of sun and shade leaves from different species of trees. 1 and 2: cross sections of two leaves from one free-standing tree of each species; 1 shows a sun leaf from the outside of the southerly exposed crown, while 2 shows a shade leaf from the inside of the crown of the same tree. 3: cross sections from shade leaves of forest trees. (Compiled from different figures in Hanson, 1917.)

This example shows that down to its finer physiological processes and its morphological properties the tree responds to differences in light and shade, and it forms both leaf activity and structure in relation to the particular illumination context in which each leaf is growing.

Tree Forms

Near The Nature Institute stands a grand white oak (*Quercus alba*). It grows at the edge of a pasture and has a trunk of about 1.2 meters in diameter (see Figure 4.6, left).

The trunk sends off long, almost horizontally growing branches that reach far out into the surroundings. The tree is broader than it is high. Not far away from this white oak—perhaps 200 meters—there is another white oak, but it looks markedly different (Figure 4.6, right). Its trunk is long and narrow and has almost no side branches up to the uppermost part of the tree. How could these two trees,

Figure 4.6. Two white oaks (*Quercus alba*): the tree on the left grows at the edge of a pasture, while the tree on the right grows in the middle of a forest; both trees grow within 200 meters of each other in upstate New York. (Sketches by Craig Holdrege; drawn to scale.)

belonging to the same species and growing in the same area, be so different from each other? The answer lies in the microenvironments (Holdrege, 2005b). The tall narrow tree grew up in the midst of other trees and is part of a forest, while the large-crowned tree grew alone at the edge of the pasture. As in the other examples I have presented so far, illumination plays a primary role in the form of the tree.

The tall narrow forest tree grew up along with other trees of about the same age; they grew out of an abandoned pasture. Together they formed a canopy that shaded themselves and each other. Lower branches died away and the main direction of growth was upward into light-filled space. In contrast, the free-standing white oak spreads its crown in all directions, in the end becoming broader than tall. In this way the relation to light and compensatory growth in response to neighboring trees has shaped the trees.

Figure 4.7 shows three trees that together form one crown: a white ash on the left, an American elm in the middle, and a smaller pignut hickory on the right. This group grows in a pasture. As in the example of the white oak, we see how the tree branches grow out into the light while growth is stunted where the trees shade themselves or each other. Since the area around the trees is open, the branches grow both outward and upward so that the three trees together form a fairly uniform crown similar to that of a free-standing, solitary tree.

Figure 4.7. A group of three trees forming a common crown in upstate New York. On the left a white ash (*Fraxinus americana*), in the middle an American elm (*Ulmus americana*), and on the right a pignut hickory (*Carya glabra*). (Sketch by Craig Holdrege.)

Plant-Environment: A Dynamic Unity

The above examples show that it is only possible to understand a characteristic of a plant—whether the shape and size of a leaf or the form and proportions of a whole tree—when one takes into account the conditions under which the plant grows. There are no characteristics without a context. Or to put it differently—every characteristic is informed by the context that made its development possible. This means that when we consider any given form or process in a plant, we need to have a kind of double awareness—an awareness of the specific qualities of the particular organism (wall lettuce in contrast, say, to gushaw) and an awareness of the environment as it is expressed through that organism. We see the environment through the plant.

Just because something is in the surroundings of an organism doesn't mean it is part of its environment (Holdrege, 1996). We can only speak of something being part of the environment of an organism when we see a response—whether in form or activity—in the organism to it. In this sense, the environment is not some "thing" or "factor" outside of the organism but is intimately bound up with it. In other words, the environment as a functional concept is inseparable from the organism (Riegner, 1993). Of course we can measure light intensity, wind velocity, soil pH, and so on, but these measurements become meaningful only when we find a concrete relation to the organism.

I have looked at plants relating to light and shade. Through such studies we can begin to gain an understanding of how a plant responds and expresses in its forms and activities the qualities of light and shade. We have seen that there are changes in macroscopic and microscopic structural characteristics and that these are accompanied by manifold physiological ones. So at all levels and in all its parts the plant responds to the conditions it grows in.

Plasticity

I have mainly been presenting examples in which I compare plants growing under two contrasting conditions. This approach allows us to find an entryway into how plants interact with the environment. In the wild, however, one can find specimens of any given species in many different microenvironments. For example, I was in New Hampshire walking along a dirt road and looked into a disturbed area bordering a meadow. I saw many specimens of wild radish (*Raphanus raphanistrum*) with their sulfur-yellow blossoms full in bloom. Within a small area—a radius of 15 meters—I found plants of all different shapes and sizes. Mainly through the flowers I could identify them as wild radish. I pressed six different specimens that show the spectrum of expressions of wild radish I found that day (Figure 4.8).

Figure 4.8. Six different flowering specimens of wild radish (Raphanus raphanistrum), collected on the same day from different microenvironments within a 15-meter radius in New Hampshire. (Silhouettes of pressed plants collected by Craig Holdrege.)

The large plant at the right was growing at the edge of a meadow while the small plant at the left was growing in an area that had been driven over by trucks many times during the previous summer. There were few plants growing in this highly compacted and disturbed ground. The other specimens were growing in between these two. Each plant developed differently in relation to the subtle differences in its microenvironment. Even in what we might call the unfavorable conditions of the highly compacted, humus-poor earth, the small specimen nonetheless developed through its full life cycle. Its growth habit changes radically—in compacted soil the plant remains small and forms few leaves, branches, and flowers—but it still unfolds as a whole plant in a way appropriate to those particular circumstances. It doesn't first try to be a large robust plant only to die in the process; rather, from the moment of germination and in the formation of each of its parts at every stage of its development, it is in living interaction—in conversation—with its environment.

So we see how a plant species can meet different microenvironments, and is able to interact with many of them, and where it can grow it brings forth a new and unique expression of its species. This shows us the plant's remarkable plasticity. The results of an experiment highlight this plasticity. Bockemühl (1973) planted field poppy seeds each month over the course of a year and then pressed the leaves from the main stem of each plant.

Figure 4.9 (following page) shows the leaf sequences of eight such plants. The date to the left of each row indicates the day on which the seed was planted, and the date on the right indicates the date the first flower opened. The leaf sequences show marked differences in size, number, and form of the leaves, as well as in the dynamics of growth. Plant 3, for example, was planted as seed in late May, formed relatively few but highly differentiated leaves, and then flowered at the end of July. When a seed was planted in mid-July (plant 5), the plant grew more slowly and developed leaves with large surfaces in a rosette; it flowered in early October. A seed planted only two weeks later also developed into a specimen with large leaves; it did not flower, and died during the winter (plant 6). When, however, the seeds were sown even later in the year (plants 7 and 8) something interesting happened—the plants overwintered in the leaf stage and developed further and completed their life cycle in the following spring. Sown late in the growing season, they grew in a way that resembles the growth of biennial plants (such as mullein) that have one season of vegetative growth and then flower in the following year.

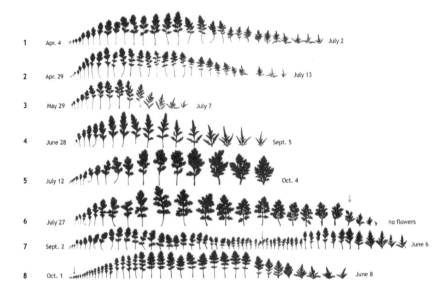

Figure 4.9. Leaf sequences (all leaves from the main stem) from eight different specimens of field poppy (*Papaver rhoeas*). The date to the left of each row indicates the day on which the seed was planted, while the date at the right of each row indicates when the plant's first flower opened. The small arrows above plants 6, 7, and 8 indicate the cessation of growth at the beginning of winter. Note that plants 7 and 8 overwintered and flowered the next summer. (After Bockemühl, 1973; reprinted with permission.)

If you only had the leaves of these different poppies, you could think you were looking at different species, so strong is the plasticity that allows them to form different shapes and sizes, and to change their growth pattern in relation to the season in which they grow. Because most of us have limited experience with plants growing in different conditions—we pay attention to what remains indicative of the species despite varying conditions, which is usually the flower—we are not cognizant enough of this remarkable plasticity in plants. It is the flexible, not-yet-determined internal constitution of the plant that allows it to meet, respond to, and interact with so many different conditions in the world.

This capacity of plants—and other organisms as well—to respond flexibly to different environmental conditions is called phenotypic plasticity (Pigliucci, 2001). There is a large body of literature documenting how plants show this plasticity in relation to a whole array of environmental conditions—both abiotic factors such as light, temperature, water, or soil pH, and biotic factors such as the density of other plant growth, herbivory by animals, or symbiotic relations with other organisms (Larcher, 2003; Pigliucci, 2001; West-Eberhardt, 2003). Phenotypic plasticity is an expression of the organism's openness to the environment during development and its ability to modify its development in relation to those conditions. The result is an organism that has formed itself by informing itself with the environment that supports it.

The Reciprocal Relationship between Plant and Environment

Up until know I have been focusing on the way the plant actively brings its environment to expression through its growth and morphology. All along we also could have been attending to how the plant is changing its environment through its own development. When a wild radish seed germinates in disturbed and compacted soil, its small roots grow down into this soil and modify it. In the following year other seeds that germinate there meet a slightly—but not insignificantly—different environment through the action of the plants in the previous year. The plants growing in a ditch that we considered at the beginning of this chapter exhibit exuberant growth. From the perspective of their environment, they are taking in more carbon dioxide and giving off more oxygen into the air than the specimens of the same species growing close by at the pasture edge. They are also bringing more water into the aboveground environment, which in part becomes integrated into their physical organism (stems and leaves consist of 70 to 90 percent water; Larcher, 2003) and in part moves through the plant, transpiring into the air and contributing to the humidity of the microenvironment. At the same time their leaves create shade, altering the environment for other plants—and for other kinds of organisms that live in that environment. As we have seen, plants growing in shady environments form larger leaves, which increase the amount of shade. Many trees growing up together in a forest shade the forest floor and create an environment that is largely influenced and sustained by the activity of the trees themselves. The roots of forest trees redistribute large amounts of water up into the soil, thereby creating an environment in which the trees and other plants can thrive—especially in times of little precipitation (Caldwell, et al., 1998; Brooks, et al., 2002). Of course plants also attract animals, providing food and habitat. And when plants die or lose their leaves, they become part of the organic compost that is transformed into soil by organisms such as fungi, bacteria, and detritus-eating animals.

Over longer periods of time, the plant-environment relation can lead to ecological succession in plant communities. When in the northeastern United States fields are no longer mowed or grazed by livestock they develop into old fields with very different species of plants. Gradually, woody plants establish and over time a forest develops. The meadow or pasture plants die off in the shady environment, and the soil gradually changes. The seeds of meadow wildflowers certainly find their way into the emerging forest, but they do not develop as they do in an open field. In the young forest different species of trees, shrubs, and wildflowers thrive than in the forest that has developed over many decades. Ongoing transformation is occurring in every plant community, which is influenced by the activity of the plants and the other organisms within the context of daily or yearly rhythms, powerful events (such as storms, droughts, or human interventions), or longer-term changes in climate.

The active nature of the organism is receiving increasing attention in ecology and in evolutionary thinking. Ecologists have coined their own term, "ecosystem engineers," to characterize the way organisms change their environments (Wright & Jones, 2006). And within the context of Gaia theory, organisms are seen in their central role as active shapers of environment and climate (Harding, 2006; Lovelock, 2000). Evolutionary biologists speak of "niche construction," indicating the process through which the organism helps to create the environment in which it lives (Laland & Sterelny, 2006; Odling-Smee, et al., 2003). Moreover, the activity of the organism itself—and not just its genes—

is seen as a shaping force within evolution (Jablonka & Lamb, 2006; Schlichting & Pigliucci, 1998).

Nonetheless, it is symptomatic of the dominance of object thinking in modern biology that despite the fact that the kinds of phenomena I just described are widely known, there has been traditionally a one-sided emphasis on organisms adapting to their environments. In the words of evolutionary biologist George Williams, "Adaptation is always asymmetrical; organisms adapt to their environment, never vice versa" (cited in Laland & Sterelny, 2006, p. 1751). In standard Neodarwinian terms, "Natural selection shapes organisms to fit preexisting environmental 'templates'" (Laland & Sterelny, 2006, p. 1751). Levins and Lewontin argue against this view in their book *The Dialectical Biologist* and describe the relation between organisms and their environment in much the same way I have:

> The activity of all living forms transforms the external world in ways that both promote and inhibit the life of organisms. Nest building, trail and boundary marking, the creation of entire habitats, as in the dam building of beavers, all increase the possibilities of life for their creators. On the other hand, the universal character of organisms is that their increase in numbers is self-limited, because they use up food and space resources. In this way the environment is a product of the organism, just as the organism is the product of the environment. The organism adapts the environment in the short term to its own needs, as for example, in nest building, but in the long term the organism must adapt to an environment that is changing, partly through the organism's own activity, in ways that are distinctive to the species. (1985, p. 69)

To make the reciprocal nature of this relation even more vivid, they point out that "the seedling is the 'environment' of the soil, in that the soil undergoes great and lasting evolutionary changes as a direct consequence of the activity of plants growing in it, and these changes in turn feed back on the organisms' condition of existence" (Levins & Lewontin, 1985, p. 134).

So the interaction of the plant with the environment is clearly a reciprocal relation. On the one hand, the plant takes the environment into itself in the development of its own organism and brings that environment to expression in its forms and functions. On the other hand, the plant is changing the environment as it develops. The plant is always contributing to the environment that supports its own life and the life of other organisms.

Context Sensitivity

"When I see a plant now, I see it as an expression of everything it interacts with, which is the whole world, and so the notion of not 'thinging' the world becomes something that happens rather than something I feel I have to try consciously to do." A participant wrote these words after a weeklong course on "the plant as a teacher of living thinking." Attending to plants can teach us to move beyond object thinking because they show us a way of life that continuously overcomes isolation and separation (the object world) by entering into a relation to the world around it, internalizing qualities of the world, and, in developing itself, bringing the other to expression. In the process, both organism and environment change.

What would it look like if our thinking and actions were modeled after the way a plant interacts with its environment? We would become more aware of how we place—or better said, plant—our ideas and actions into concrete situations in life. We would become flexible enough to actively adapt to what a situation brings toward us, knowing that the fruitfulness of our thoughts and work depends on those very circumstances. We would gain an ever-greater sensitivity to the context that informs any given situation, problem, or challenge we perceive. In other words, just as we see the individual form of a plant or a leaf as an expression of its environmental relations, our viewing and questioning would not see isolated things, problems, or obstacles, but would inquire into the context that informs them. And we would become aware of how the way we have thought and acted has affected the context we are interacting with.

The study of the plant is a training ground for contextual and living thinking and also presents us with a kind of ideal of living, context-sensitive interaction. Clearly, we have much work to do to put ourselves, cognitively, in a contextual relation to the world that is similar to the way the plant grows in its world. To what extent can we do willfully—in thought and action—what the plant lives?

Relational Reality

Inasmuch as the plant in its form, structures, and processes reveals the world it has grown in, and the world carries the imprint of the plant, I cannot separate them. I can make distinctions but I cannot divide, since the plant reveals to me not only itself as a species but also a larger context (in which, of course, many other species participate, each in their own unique ways). The reality is "organism-environment." I see the one through the other.

In a living view of the world we are always dealing with relational reality that is precisely no thing but not nothing. Because our language is full of nouns, and thingness provides clarity of mind, there's not much getting around speaking of entities ("an organism," "this plant," etc.) in order to communicate. What is essential is that we keep our attention focused on *what we actually mean*. And what we mean is that each thing or entity is a unique and dynamic focus of relations, and it is what it is by virtue of these relations.

In this sense each thing is both itself and other than itself. It has distinctness and yet is not separate. This seems paradoxical, but it is clearly what the plant shows us. And there is a place very close to home where we know the relational nature of life to be true and where it is completely clear that entities can be distinct and yet grow only by virtue of each other. That is in human conversation, or dialogue.

What distinguishes conversation, or dialogue (I use the terms interchangeably), from a monologue, a discussion, or a debate? In a conversation I listen and I am not particularly invested in convincing you as my partner of my point of view. I make the effort to understand what you are intending, meaning, and saying. But I don't only take in what you say, I respond, perhaps by raising a question or commenting on what you have said; I might bring in a new perspective that helps the dialogue to move along. There is a back and forth, a taking in and giving. Your thoughts have been taken up by me and vice versa. What lives in me is also alive in you. We are not separate from each other; a merging of minds has occurred. Although I can still say "I am myself," I can also say that I have grown by virtue of my interaction with you. I have you in me. While it is certainly true that each of us goes our own way physically after a conversation, a kind of mental symbiosis has formed that both of us can still participate in, and further dialogue can lead to new growth.

So I am both myself and other than myself. The plant is both itself and other than itself. Everywhere we look in the world we find centers of activity embedded in fields of interaction. This is what life shows us. That doesn't make it easier for us modern human beings to fully acknowledge the relational reality of every thing, process, or situation. And it is even more challenging for us to realize that these relations always include us; we don't stand outside them.

Participation

Let's begin where matters are pretty straightforward. Without plants we would have no oxygen to breathe and no food to eat. Our existence on earth is wholly dependent on plants. We need to take in and transform what plants produce. There are countless ways we interact with plants—they provide environments for us and we actively alter them and their environments, as when we cut down a forest, plant a garden, or do experiments that vary light or soil conditions that plants are growing in. These physical interactions can be very subtle, as when researchers find that touching plants they are marking in the field has an effect on the outcome of experiments (Niesenbaum, et al., 2006).

We interact with plants in other ways as well: We smell the scent of pines on a hot sunny day, we enjoy the radiant flower of a poppy, or we bask in the cool green of a beech woods. Here we turn as sentient beings toward plants, and they enrich us and become part of us.

We also turn to the plant with questions and interest. We study the way plants grow and ponder the manifold variety of their forms. In this book we have gained insight into patterns in plant development, into the plant's dynamic adaptability, and into its relational way of being. Surely, like the joy we experience through plants, such insights enrich us. But that is not the whole story.

It is also the case that, through these efforts, qualities of the plant are being recognized and articulated by us. This means that the plant shows a new expression of itself. Just as a plant that has grown in a particularly wet spring may manifest characteristics that it has not shown before in other conditions, the human perceiver-thinker provides an environment for the plant through which it can reveal aspects of itself that otherwise would not come to appearance. As Steiner puts it:

> Plant a seed in the earth. It puts forth root and stem; it unfolds into leaves and blossoms. Place the plant before yourself. It connects itself, in your mind, with a definite concept. Why should this concept belong any less to the whole plant than leaf and blossom? You say the leaves and blossoms exist quite apart from a perceiving subject, but the concept appears only when a human being confronts the plant. Quite so. But leaves and blossoms also appear on the plant only if there is soil in which the seed can be planted, and light and air in which the leaves and blossoms can unfold. Just so the concept of a plant arises when a thinking consciousness approaches the plant. (2006, pp. 65–66; translation slightly modified by Craig Holdrege)

In other words, the understanding we gain through the study of plants is part of the plant's relational reality. When a plant has been noticed and studied, when aspects of its life have lit up in a human mind, an otherwise dormant potential in the plant comes to expression in the new environment that we provide through human inquiry.

I know this sounds strange. What I am trying to do is to articulate how we as human beings participate in a unique way in relational reality. I want to acknowledge the activity of human knowing as part of the ecology of the earth, and not treat it as some otherworldly capacity. Plants have the miraculous ability to form organic substances out of their interactions with light, air, and soil. We don't have that ability, but we have a different ability: we can inquire into the nature of things, study nature for its own sake, and try to get to know better the world we are part of. In this interaction we establish new relations and new aspects of the world come into appearance.

As knowers we are evolving beings. We don't have total insight. Our understandings are imperfect. You could say that we are involved in the process of enlivening the soil of the mind so that what we interact with can come better to expression. Many of our frameworks and models may be for the living world something like growing in a very thin layer of poor soil, but through continued work to adapt our sensibilities to what we are interacting with, this soil can become richer. I don't want to stretch the metaphor, but it points in the right direction.

The realization of the depth of our participation in the world is both freeing and humbling. We become aware that we are not aliens on this planet, even though much of the time we think and act as if we were. And since—whether we like it or not—we are bound up with everything else, and not just in physical terms but as knowing beings as well, we can choose how to shape our engagement. To what degree do we work on enhancing the quality of our participation, or do we continue to blindly tear into the fabric of things through our set ways of knowing and doing?

The enhancement of our manner of engagement is what I am concerned with in this book: to learn to listen and to find ways for the creatures of this world to speak ever more distinctly through us. To let the other speak, but nonetheless to be an active participant, is what happens, as we just saw, in a dialogue or conversation. This is the quality that can inform all of our interactions—whether they involve studying natural phenomena, developing a new technology, teaching children, or working in ecological restoration. In what follows I consider how dialogic inquiry within science can unfold and discuss the ever-present tension between exploration and theory in science.

Dialogic Inquiry

When in the exercise of our powers of observation we human beings undertake to confront the world of nature, we will at first experience a tremendous compulsion to bring what we find there under our control. Before long, however, these objects will thrust themselves upon us with such force that we, in turn, must feel the obligation to acknowledge their power and pay homage to their effects. When this mutual interaction becomes evident we will make a discovery which, in a double sense, is limitless; among the objects we will find many different forms of existence and modes of change, a variety of relationships livingly interwoven; in ourselves, on the other hand, a potential for infinite growth through constant adaptation of our sensibilities and judgment to new ways of acquiring knowledge and responding with action. (Goethe, 1995, p. 61 (transl. modified by Craig Holdrege); written 1807; published 1817)

When Goethe speaks of the "potential for infinite growth through constant adaptation of our sensibilities and judgment to new ways of acquiring knowledge and responding with action," he is describing the plasticity of the human mind. Our untapped potentials grow into reality through interaction with the world we encounter. We engage in the "many different forms of existence and modes of change" through which we become aware of a "variety of relationships livingly interwoven." This interaction stimulates growth in us as knowers, and the more we grow, the more of the "limitless" nature of the world becomes apparent and also finds expression through us.

But the "tremendous compulsion to bring what we find under our control" works against such a dynamic and living way of knowing. When we "confront" nature, we see it as something outside of and foreign to us and want to bring it "under our control." We are more interested in bringing it into our sphere than in listening to what is has to say. In respect to scientific knowledge, philosopher Immanuel Kant argued that this stance is necessary for the fruitful progress of human knowledge. He states that while we do want to be taught by nature, we "must not, however, do so in the character of a pupil who listens to everything that the teacher chooses to say, but of an appointed judge who compels the witnesses to answer questions which he has himself formulated" (Kant 1965, p. 20; written in 1787). In this mode of inquiry, we are in control, since we set the terms of the interrogation.

Goethe experienced this desire for control in himself, but he also experienced that the phenomena "thrust themselves upon us with such force that we, in turn, must feel the obligation to acknowledge their power and pay homage to their effects." Here Goethe points to the necessity of remaining open to the phenomena, to a keen perceptual sensitivity that counterbalances the desire to control.

Delicate Empiricism

Goethe not only studied plants (and animals) in great detail, he also carried out extensive research into light and color. He performed countless observations and experiments, continually looking at the phenomena from different points of view and in experiments varying the conditions in new ways (Goethe, 1995, chapter 7). Even a critic of Goethe's approach would have to grudgingly admire the persistence with which he returned again and again to observation and never tired of looking at phenomena from new angles.

Why did he do this? First, he noticed immediately when he began to study color phenomena that they are highly dependent on the conditions under which they are observed. He saw that changing conditions, for example, in light intensity, the distance and angle of view, or in the state of the observer, all contribute to the phenomena of color. Just as you can't speak of a plant separately from its environment, you can't study colors as isolated facts. Therefore, Goethe realized, to do justice to color phenomena, he would have to study them within a variety of contexts.

Secondly, he was both aware and wary of our human propensity to form judgments about things based upon few observations or experiments. As he writes in his seminal essay, written in 1792, "The Experiment as Mediator of Object and Subject":

> We cannot take great enough care when making inferences based on experiments. We should not try through experiments to direct-ly prove something or to confirm a theory. For at this pass—the transition from experience to judgment, from knowledge to appli-cation—lie in wait all our inner enemies: imaginative powers that lift us on their wings into heights while letting us believe we have our feet firmly on the ground, impatience, haste, self-satisfaction, rigidity, thought forms, preconceived opinions, lassitude, frivolity, and fickleness. This horde and all its followers lie in ambush and suddenly attack both the active observer and the quiet one who seems so well secured against all passions. (2010, p. 20)

Faithfulness to the phenomena keeps our judgments grounded and world-related. We become cautious and circumspect and do not get carried away by the levity of a grand idea. Goethe came to the conclu-sion that "one experiment, and even several of them, does not prove anything and that nothing is more dangerous than wanting to prove a thesis directly by means of an experiment" (2010, p. 20). By varying the experimental conditions and by looking at things from different an-gles, we gain a comprehensive and differentiated picture that helps us leave behind all-too-narrow, schematic, or rigid conceptions. There-fore, in Goethe's view, "We accomplish most when we never tire in exploring and working through a single experience or experiment by investigating it from all sides and in all its modifications" (2010, p. 22).

Goethe was motivated by his love of the phenomena and his desire to do justice to them in scientific explorations. Scientists should work to "take the measure for knowledge—the data that form the basis for judgment—not out of themselves but out of the circle of what they observe" (2010, p. 19). Toward the end of his life Goethe called this approach "a delicate empiricism that makes itself utterly identical with the object" (1995, p. 307). We become delicate empiricists when our primary goal is to let the phenomena speak and therefore explore them from many sides and remain highly conscious about the way we interact with them through our ideas. In so doing we work to remain aware and critical of our tendency to overpower the phenomena (that is, control them) with our interpretations, models or theories:

> We can notice that a good mind is all the more artful the less data lie before it. To show its command, it selects a few flattering favorites from all the available data and knows how to order what is left over to show no contradictions. It knows how to confound, enwrap, and push aside the opposing data, and in the end the result resembles a despotic kingdom rather than a freely organized republic. (2010, p. 22)

In all scientific inquiry there is a tension between exploration—the search for new observations and the uncovering of new phenomena—and the desire to understand, to find patterns and lawfulness in the world. As Goethe writes:

> We take pleasure in a thing in so far as we form an idea of it and when it fits into our way of looking. We may try to raise our mode of thought so far above the everyday mode as possible and strive to purify it, but nonetheless it usually still remains only a mode of thought. It follows that we attempt to bring many phenomena into a certain graspable relation to one another that they may, looked at more closely, not have. And we have the tendency to form hypotheses and theories and to craft terminology and systems accordingly. We cannot condemn these attempts since they arise with necessity out of the organization of our being. (2010, p. 20)

Nonetheless, we can work to move beyond such limitations. Delicate empiricism is a method to overcome the inertia of the human mind that remains within already formed paradigms and well-trodden mental pathways. To transcend deep-seated habits of thought in the search for more adequate insights requires an arduous transformation of human capacities—and this is a lifelong process. For this reason Goethe added to his words about delicate empiricism that such "enhancement of our mental powers belongs to a highly evolved age" (1995, p. 307). Along the way we can make use of hypotheses and conceptual frameworks. What is important is how we use them (Zajonc, 1999).

Goethe is critical of hypotheses because they constrict: "All hypotheses hinder us in beholding and considering the objects and phenomena from all sides" (1973, p. 441). Therefore it is important to gain a free and sovereign relation to one's hypotheses: "Hypotheses are the scaffolding that one erects in front of a building and that are dismantled when the building is completed. To the worker the scaffolding is indispensable, but he must not take it for the building itself" (p. 441). As Goethe remarks, it is a freeing experience to cast off a way of seeing and to venture into a new one:

> When we free our mind from a hypothesis that unnecessarily limited us, that compelled us to see falsely or half-way and to combine falsely, that compelled us to brood rather than to look, to commit sophistry rather than to discern, then we have performed a great service. For then we can see the phenomena freely and in other relations and connections; we can order them in our own way and have occasion to err in our own way—an invaluable opportunity as long as we are able to understand our own errors. (Goethe, 1973, p. 441)

Key here is that scientists remain aware of how their concepts are at work in the process of coming to understand. Without this awareness we run the danger of conflating our ideas with the phenomena themselves, not realizing that we are speaking of a phenomenon only in the restricted terms of a specific hypothesis, framework or theory. This is especially crucial as a healthy deterrent to the widespread tendency to make universal claims, which is essentially a form of unwarranted extrapolation and reveals lack of context sensitivity, as well as a mind all-too-confident in its powers of generalization. Rudolf Steiner addresses precisely this problem in relation to the so-called laws of nature:

> You can find, for example, in physics books the law of imperme-ability presented as an axiom: at a point in space where one body is, another cannot be at the same time. This is stated as a general characteristic of bodies. However, what we should say is that those bodies or beings are impermeable whose nature is such that where they are in space, no other being of the same nature can be at the same time. We should use concepts only to distinguish one area from another. We should only set up postulates and not give defini-tions that claim to be universal. We should not declare a law of the conservation of energy and matter, but rather seek the beings for which this law has meaning. (1996b, p. 78; translation modified by Craig Holdrege)

This is a wonderful example of wakeful, concrete, and context-sensitive concept formation. It makes vividly clear that the question is not whether we should have ideas that guide our investigations. Rather, the question is, what is the quality of those ideas? Do we apply them consciously and use them as ways of illuminating a phe-nomenon from a certain perspective? Can we disengage ourselves from an existing theory or hypothesis so that the phenomena can re-veal new sides and dimensions? Are we ready to let the encounter be the stimulus for our ideas to grow and transform, and maybe even then to fall away and dissolve? The vitality and depth of a scientific idea would show itself in the way it grows and transforms in the course of the dialogue with the phenomena.

Darwin: Exploration and Theory

The work of Charles Darwin highlights in a vivid way the relation between exploration and theory formation in science. As a young man Charles Darwin was an avid naturalist (Desmond, &

Moore, 1994). He loved to observe and collect. He was able to live out his passion when at the age of twenty-two he became the naturalist for the *Beagle*, a ship that sailed around the globe over the course of five years (1831–1836). During the journey on the *Beagle* Darwin collected, among other things, many fossils and began to build up a picture of how landscapes had formed through geological time. He observed and collected plants and animals in different environments and noticed how they were exquisitely adapted to the conditions in which they lived. He had an eye for the subtle differences between specimens of a species that others might easily overlook. He saw, wherever he looked, great variation in nature, and, in the years that followed his return to England, he came to the conclusion that there are no fixed boundaries between species. One could say that, guided by the immersion in the natural world itself, Darwin began to form an idea of the transformational nature of life on earth—not only within the individual but also within populations, species, and landscapes. For Darwin, nature was characterized by variation, change, and flux, and having gained this view, he saw himself leaving behind the more static and dominant view of his time that species are and always were distinct entities. The young Darwin was working in an exploratory Goethean spirit and contributed to a living, dynamic, and ecologically informed view of life on earth.

Throughout his long life Darwin never tired of observing and pondering the significance of natural phenomena, whether it was the contribution of earthworms to soil formation, the movements of plants, or the biology of barnacles. But Darwin was not satisfied to build up a dynamic picture of life on earth. He wanted to explain how organisms evolved. What did "explain" mean to him? It meant finding a general theory—a causal mechanism—that could show how species transform and become so well adapted to their environments. He was compelled by the quest to find one general, fundamental, and logical idea—an urge that is so characteristic of modern science and that culminates in the desire to find the one theory of everything. During his life, Darwin's earlier exploratory manner of research became increasingly overlaid by a theory-driven one.

In formulating his theory of evolution through natural selection, Darwin was strongly influenced by Thomas Malthus' *Essay on the Principle of Population* (written in 1798). Malthus argued that human population has a tendency to grow that is "indefinitely greater than the power in the earth to produce subsistence for man" (1999, p. 13). In his view, illness, wars, famines, and natural disasters become important factors in controlling population growth and thereby contribute to the survival of humanity. Darwin wrote in his seminal book *Origin of Species* (first published in 1859) that his own theory is "the doctrine of Malthus applied with manifold force to the whole animal and vegetable kingdoms" (1979, p. 117). This is a classic statement of a theory-driven approach: one finds an idea, generalizes it, and then applies it far beyond the boundaries within which the idea was originally conceived.

It was important for Darwin that the basic tenets of his theory seemed to follow from facts of nature. Since he had extensive knowledge of the natural world, he was well aware of the immense challenge of finding a theory that did, to his mind, justice to all the facts. He brought together in his mind three areas that were based on thorough observation: First, there is heritable variation within species and populations. Second, there are many more offspring produced by plants and animals than actually survive. Third, plants and animals are well adapted to their environments. He then connected these phenomena in thought: Since not all offspring survive, it makes sense that those that do survive will be those that are better adapted to their environment than those that perish. He spoke of natural selection as the "principle by which each slight variation, if useful, is preserved" (1979, p. 115).

When well-adapted variations produce fertile offspring, over time they become the dominant form of the species. With enough time, in his view, the species will evolve, provided that new variations are being produced that are better adapted to existing conditions, or that are well adapted to a changed environment. Evolution occurs as the cumulative effect of many small changes in organisms as a result of the "struggle for existence" (1979, chapter 3) over long periods of time. This is a compelling train of thought and once it became clear in his mind, Darwin proceeded to interpret all biological phenomena in its light.

Remarkably, with a keen eye for his own thought processes, near the end of his life Darwin noted how his mind had changed as a result of his exclusive concern with the theory of evolution through natural selection. Looking back to what he had written in his journal during the *Beagle* voyage, he remarked:

> In my Journal I wrote that whilst standing in the midst of the grandeur of a Brazilian forest, 'it is not possible to give an adequate idea of the higher feelings of wonder, admiration, and devotion which fill and elevate the mind.' I well remember my conviction that there is more in man than the mere breath of his body. But now the grandest scenes would not cause any such convictions and feelings to rise in my mind. It may be truly said that I am like a man who has become colour-blind, and the universal belief by men of the existence of redness makes my present loss of perception of not the least value as evidence. (Darwin, 2005, p. 76)

In other words, his worldview had taken on firm contours, and in the wake of the development of his theory the feelings of wonder and reverence he previously felt had dissipated. In the end, he could see the natural world only through the lens of his theory and believed that what he saw through this lens was true, and that, more importantly, it was all that was needed to explain nature—nature contains nothing or is expressive of nothing beyond what his theory encompasses. For this reason, he interprets the loss of his earlier feelings of wonder and devotion as the loss of delusory feelings toward something higher or deeper in nature—the divine, or God. Everyone else in the world may claim he is color blind, but that in no way sways him. Darwin observed that "my mind seems to have become a kind of machine for grinding general laws out of large collections of facts" (2005, p. 113). These general laws, for him, are the laws that govern the natural world.

Today Darwin's theory (of course modified and refined by over a century of further scientific work) has become, in Thomas Kuhn's sense, "normal science" (Kuhn, 1996, Chapter 3). It is the generally accepted paradigm. All biology students learn to interpret the living world through the lens of natural selection. Whatever biological characteristic one confronts, the question is essentially the same: how does this characteristic contribute to the survival of the species? If I can discover how it increases or at least does not hinder the fitness of the organism, I believe I have understood it. It is quite easy to come up with theoretical claims about how it is beneficial for a bird to have just the wing structure it does, how the highly developed forelegs of a mole allow it to live the life of a burrowing animal, or how the long neck of the giraffe lets it feed on leaves higher up in trees than are accessible to most other four-legged animals. After all, animals and plants that exist are well adapted and all their characteristics do in one way or another allow their survival.

But it is another matter to try to explain the evolutionary process— to formulate the set of causal relations—that brought forth such adaptations. For example, about giraffe evolution Darwin writes (in the sixth edition [1872] of *Origin of Species*):

The giraffe, by its lofty stature, much elongated neck, fore-legs, head and tongue, has its whole frame beautifully adapted for browsing on the higher branches of trees. It can thus obtain food beyond the reach of the other Ungulata or hoofed animals inhabiting the same country; and this must be a great advantage to it during dearths.... So under nature with the nascent giraffe the individuals which were the highest browsers, and were able during dearth to reach even an inch or two above the others, will often have been preserved; for they will have roamed over the whole country in search of food.... Those individuals which had some one part or several parts of their bodies rather more elongated than usual, would generally have survived. These will have intercrossed and left offspring, either inheriting the same bodily peculiarities, or with a tendency to vary again in the same manner;

whilst the individuals, less favoured in the same respects will have been the most liable to perish.... By this process long-continued, which exactly corresponds with what I have called unconscious selection by man, combined no doubt in a most important manner with the inherited effects of the increased use of parts, it seems to me almost certain that an ordinary hoofed quadruped might be converted into a giraffe. (pp. 177 ff.)

With this description Darwin paints a picture of how the giraffe could have evolved its long neck. It makes sense and seems logical. It is a typical example of how characteristics are explained within a Darwinian framework. But this kind of explanation is actually, as Gould and Lewontin (1979) call it, an "adaptive story." It may or may not be true; there may be alternative stories that are just as logical and compelling. For example, biologists have come up with other adaptive stories that "explain" the giraffe's long neck, such as thermoregulation or competition among males (see Holdrege, 2005a). I think you would be hard-pressed to find any characteristic for which one could not find at least different compelling adaptive stories. As Gould and Lewontin (1979) point out, it is all too easy to fall into the trap of confusing "the fact that a structure is used in some way with the primary evolutionary reason for its existence" (p. 587). Because a long-necked giraffe does occasionally reach high into trees to feed (usually at times when there is adequate food supply!), does not mean this is the evolutionary reason for its existence. Such conflation is a form of self-deception that the theoretical mind so easily succumbs to in its search for one-size-fits-all explanations.

The more you study an animal—or any organism—in detail, the less convincing adaptive stories become. They pale before the richness of the whole organism itself. It is fascinating to follow how Darwin unfolded his theory and to read how he tries to apply it to phenomenon after phenomenon. In the process, the abstract notion, once formed and consolidated, tends to take on a life of its own—revealing everywhere the same "explanation" only because it adamantly remains true to itself and takes into account just those aspects of the phenomena that fit into the framework. This is when theory runs the danger of becoming the kind of "lethal generality" Goethe so hoped science based on a delicate empiricism would avoid (1995, p. 61).

For example, Darwin writes, "Nature cares nothing for appearances, except in so far as they may be useful to any being" (1979, p. 132). This follows from his basic premise; if his theory is right, then this must be the case. But this exclusively utilitarian point of view also radically limits the way one is "allowed" to judge organic forms. When you free yourself from this bias, enter a more exploratory mode, and allow the phenomena to once again become your primary teacher, the appearances may begin to show new and unexpected sides. For example, the pioneering studies of Portmann (1967) and Schad (1977), as well as the more recent work of Lockley (2007) and Riegner (2008), demonstrate that there are underlying nonadaptive morphological patterns in body form, proportions, and appearance (e.g., in bird plumage).

Although there have been—both before and after Darwin—biologists and evolutionary thinkers who have interpreted evolution from nonadaptationist perspectives (see Holdrege, 2009, for a partial bibliography), these ideas have remained decidedly on the margin and have been largely ignored by the "normal science" of mainstream biology and biology education. What biology student today would be encouraged to come up with a different approach to evolution from the one that forms the fabric of all the textbooks and courses? Generally speaking, the "facts" of science are taught without the teacher or professor making the conceptual framework explicit or providing students with the opportunity to look at the same facts from other points of view. The student learns within the cloak of a theoretical framework and is never taught to take a look at that cloak or given occasion to remove it and try on a different one. When a way of viewing the phenomena becomes the unreflected bias of all investigation then, while certainly leading to knowledge within that limited framework, it limits insight in a broader sense. It limits by cultivating only one kind of approach, and it limits by marginalizing other perspectives.

Darwin's theory of evolution is an example of a powerful and widely acknowledged theory. It has stimulated vast amounts of research. At the same time it has decidedly limited the scope of scientific inquiry through its dominance in the scientific establishment. Inasmuch as scientists attempt to apply a theory to every phenomenon it becomes more and more general and abstract. You might recall Ken Wilber's remarks in relation to systems theory that I quoted earlier: "Precisely in its claim and desire to cover *all* systems [it] necessarily covers the least common denominator, and thus nothing gets into systems theory that, to borrow a line from Swift, does not also cover the weakest noodle" (Wilber, 2000, p. 122). Therefore, he concludes, a general theory may be "fundamental," but it is also the "*least* interesting, least significant" (p. 122; his emphasis), since it only takes into account what fits within the theory. When I state that the scarlet-red carapace of a milkweed beetle, the deep blue plumage of a bluebird, and the white fur of an arctic fox are all related to survival, I am in a sense saying something fundamental, but I am also saying nothing specific about those particular animals. When characteristics become no more and no less than survival strategies, then they have become little more than shades of gray in the limited color spectrum of the general theory. But that is not their fuller reality.

Evolving Knowledge

Because a theory is always a limited human notion, it cannot do justice to the complex nature of the world. A general theory, therefore, is essentially something to be overcome as human understanding and science evolve. The more tentatively guiding concepts are held, the more dynamic and exploratory the mode of inquiry becomes. In dialogue with the phenomenon a new quality of knowing can emerge—the perceptual-conceptual beholding of essential relationships and patterns. These are not abstract ideas, and they are also not fixed. They evolve with further inquiry as the human being evolves. When this happens, one can rightfully speak of an organic way of knowing. Knowledge grows out of the careful interaction of human being and phenomenon in the way the plant develops out of the interaction of plant species with its environment. Flexibility of mind, openness to the new, and the ability to let each new phenomenon stimulate the growth of fresh conceptions are the plant-like qualities that characterize a living and evolving science.

CHAPTER 5

The Story of an Organism

All I am saying is that there is also drama in every bush, if you can see it. When enough men know this, we need fear no indifference to the welfare of bushes, or birds, or soil, or trees. We shall then have no need of the word "conservation," for we shall have the thing itself.

Aldo Leopold

ONLY WHEN WE GET to know the life and ecology of the organisms that are part of our world do we have the possibility of seeing "drama in every bush." Such learning can happen in a variety of ways, but nothing can substitute for firsthand experience of the life history of another organism. To enter that world, to see its intricacies, and to realize that in every direction you look you will find new connections and relationships—this is a central foundation of sustainability. For how are we to care for what we don't know?

The kind of knowing I mean here is knowing based in experience, in meeting the world we inhabit. When we experience we are engaged as whole human beings—we observe, we ponder, we react, we intervene, our feelings are aroused. This kind of knowing is anything but abstract, and it is not the kind of fear-arousing knowledge that informs so much of the environmental debate today. Aldo Leopold captures the essence of the problem and the task when he writes, "I have no hope for conservation born of fear. The 4-H boy who becomes curious about why red pines need more acid than white [pines] is closer to conservation than he who writes a prize essay on the dangers of timber famine" (1999, p. 165). It is engaging in concrete phenomena that gives us a relation to them that goes beyond mere information. As Leopold puts it, "We can only be ethical in relation to something we can see, feel, understand, love, or otherwise have faith in" (1989, p. 214).

The Importance of Story

Environmental historian William Cronon writes about the importance of stories—narratives—in environmental history studies. Without stories we have a mere assemblage of fact; in history this is a chronicle. "The principal difference between a chronicle and a narrative is that a good story makes us care about its subject in a way that a chronicle does not" (Cronon, 1992, p. 1374). The story is not something "made up," nor does it meet the criterion of the television detective Joe Friday's admonition, "Just the facts, ma'am." If we try artificially to have just the facts (a list of events or a compilation of an organism's traits), "It becomes hard to evaluate the relative significance of events. Things seem less connected to each other, and it becomes unclear how all this stuff relates to us" (p. 1351). "Stripped of the story, we lose track of understanding itself" (p. 1369). Cronon concludes that

> the special task of environmental history is to assert that stories about the past are better, all other things being equal, if they increase our attention to nature and the place of people within it. They succeed when they make us look [for example] at the grasslands and their peoples in a new way.... As Aristotle reminded us so long ago, narrative is among our most powerful ways of encountering the world, judging our actions within it, and learning to care about its many meanings. (p. 1375)

Clearly, a scientific narrative will differ in many ways from an environmental history that has human action as a central focus. But it is still a narrative. Most scientific narratives of plants and animals today are written from a Darwinian perspective, which provides an explanatory framework that gives genetic variation, differential reproduction, and natural selection as the basic elements of the story line underlying every particular story. The plot is always the same, but the characters are different. In addition, most Darwinian stories are usually not stories of the whole organism, but rather of one aspect or part of its anatomy, physiology, or behavior.

I do not impose an explanatory framework onto my narrative. Rather, I seek to build up a picture of an organism as a whole, which plays a large part in telling its own story. I have carried out a number of such whole organism studies.[1] In each of these cases my attempt was to portray these organisms as the unique creatures they are. Each of these organisms, for one reason or another, caught my interest and a detailed study ensued. The mutual learning that occurred during the plant study in one particular summer course led me down the pathway to yet another organism.

The Story of Common Milkweed

I had casually observed common milkweed (*Asclepias syriaca*, Asclepiadaceae) but never really paid too much attention to it. True, I was fascinated by its big globes of flowers and, in the fall, by its beautiful seeds that float through the air on their tufts of white silk. I also knew that common milkweed is the main food plant for monarch butterfly larvae. Soon before one summer course started, the flowers of common milkweed began to open and I looked at them for the first time more closely. I realized that the plant has a highly complex flower structure and, in addition, I observed how the flowers were being visited by many different insects. Milkweed had finally really caught my attention, and I decided that we should focus on it for our initial plant study in that weeklong course.

This study proved to be particularly intense. Milkweed drew us all into its world of refined structures. It took us a good deal of time just to get clear about the flower parts and their relation to more "typical" flowers. (There were a number of trained biologists in the course.) We also observed interaction with insects and saw how flies sometimes became caught in the flowers and died. After four morning sessions dedicated to milkweed, we had made an acquaintance with the plant.

And we had begun to get a glimpse of its unique characteristics, which became all the more apparent through a comparison with St. John's Wort (*Hypericum perforatum*), which was also flowering at the time. During these days we had begun to see milkweed as a singular composition expressing itself, on the one hand, in robust structures such as the rhizomes, upright stems, and leaves and, on the other hand, in the high-grade refinement of the flower. This glimpse of the nature of common milkweed initiated my journey to get to know the plant better. It was a wonderful gift to have a study initiated by a group effort during a course.

My attempt here is to give a portrayal of common milkweed—its life history and relations to some of the organisms with which its life is bound up. The chapter integrates my own observations with information taken from the scientific literature. This literature is a reflection of the countless hours researchers have spent observing common milkweed and its ecology, designing and carrying out experiments, and interpreting their findings.

In a whole-organism study the task is to bring together one's own observations with those of many other researchers and to paint a coherent picture of the plant and its relationships. In Goethe's words, the goal is "to portray rather than explain" (1995, p. 57). Such a portrayal is a kind of story of the plant. To craft such a portrayal, it is necessary to omit many of the hypotheses and interpretations that presently guide most of the scientific research on the plant. Such hypotheses often intend to find an explanation of why, for example, milkweed has its specific flower form, why it secretes so much nectar, or why it contains in its leaves and stems a sticky, toxic sap. My question is not "why?" but rather "how?" I want to see how milkweed is formed, how its parts relate to each other, and how it relates to its environment.

I hope this portrayal will provide a vivid picture of a remarkable plant and at the same time serve as an example of a process that can be applied to other organisms as well.

Vegetative Development

It is early May in upstate New York. When you look onto roadsides and old fields there is still much of the past to be seen—the dead brown and gray remnants of last year's growth. The dry and brittle leaves of some grasses and wildflowers, and the crisscross of matted stems from asters, goldenrods, and milkweed provide the immediate surroundings for this year's fresh green emerging vegetation. The shoots of grasses, goldenrods, and other wildflowers are rising up out of the soil; common milkweed shoots emerge only later. You have to get down close to the ground to see the stout little green and reddish-brown spears growing through the leaf and stem litter (see Figure 5.1).

And where you find one shoot emerging, you usually find many. Common milkweed grows in colonies that tend to get more populated each year, unless something in the environment inhibits their exuberant growth. There can be hundreds or thousands of shoots,

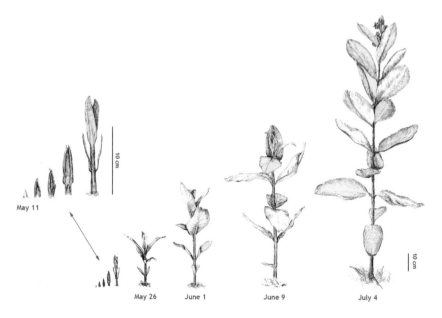

Figure 5.1. Shoot development in common milkweed (*Asclepias syriaca*). Shoots from one colony, drawn to scale. Summer 2007. (Drawings by Craig Holdrege.)

depending on the size of the colony. All these little shoots grow from buds that have overwintered; they developed the previous summer on rhizomes (underground stems) and remain dormant until the following spring. The rhizome grows and branches each year—never showing itself above ground—forming an extensive network out of which the many individual shoots grow (Figure 5.2).

A colony is thus one large plant. You could compare it with an individual tree, but instead of growing a woody trunk that extends upward and branches outward to create a lasting form, common

Figure 5.2. Rhizome of a common milkweed (*Asclepias syriaca*) laid bare during the late spring. Top: branching rhizome; dime at the right indicates scale. Bottom: close-up.
a: rhizome; b: roots; c: shoot from previous year; d: buds that were formed in the summer/fall, some of which will unfold in the following spring; e: scars from the base of a shoot from the previous year. (Photos Craig Holdrege.)

milkweed branches underground and some old parts of the rhizome die away while new ones arise. Botanically speaking, a colony is a clone—the individual shoots are genetically identical, having originated from a single seed through vegetative growth. So when I observe what looks like an individual specimen of common milkweed, the shoot I am observing is one side branch of a much larger plant, namely the whole colony.

One perceptual hint of this larger unity is that milkweed colonies are usually quite uniform and differ from other colonies in characteristic ways. One colony might exhibit shoots that are long, narrow, and distinctly red-tinged; another colony may have more pointed-tipped leaves; a third may have especially deep-pink-colored flowers. Such characteristics are hereditarily anchored and therefore are common to all parts of the colony.

Back to the emerging shoots. I will focus now on one shoot and its development, but we should keep in mind that what I'm describing is occurring with many other shoots within the colony at more or less the same time. As a shoot grows upward, the first small lance-shaped leaves begin to unfold. The leaves are arranged in an opposite pattern, meaning that two leaves emerge at the same height out of the stem, opposite each other. The next pair of leaves is offset by about ninety degrees so that a distinct leaf arrangement (phyllotaxis) emerges (Figure 5.1).

The stem is stout and grows quite straight upward. It is significantly thicker than its counterpart in asters or goldenrod, but not so dense. While the first leaf pairs near the ground are small, the leaves soon became long and broad. Milkweed has much bigger leaves than other plants growing in the roadside/old-field community. The leaves have a short thick petiole that becomes the distinct midrib of the leaf blade. The midrib branches off into fairly straight, diagonally ordered side veins that are clearly visible on the leaf underside. The oblong leaves have a smooth, clear margin (Figure 5.3).

Over the course of May and through June the shoot grows vigorously, reaching a height of about 1 to 1.5 meters. It surpasses in height the goldenrods that had begun their development a few weeks before

Figure 5.3. Pressed leaves from one shoot of common milkweed (*Asclepias syriaca*). The first leaves are at the bottom, the last leaves are at the top. When the plant is in full flower, the lower pairs of leaves have died and fallen off the plant. The lower pairs were picked and pressed soon before they had wilted. (Craig Holdrege.)

the milkweed. As the shoot extends, new leaf pairs unfold, each pair larger than its predecessor. After a shoot has developed around seven to nine pairs of leaves, the first flower buds become visible among the still unfolding uppermost leaves. The stout little stems (peduncles) that carry the groups of flower buds do not grow out of leaf axils, as is the case in most flowering plants. Rather, in all but one species of the genus *Asclepias*, the flowers grow out of the main shoot slightly to the side of a leaf (Woodson, 1954). Interestingly, where flowers develop, the leaf arrangement also changes. Subsequent leaf pairs are no longer perpendicular to each other, but shift to an angle of about 120 degrees. Also—and this is typical in most flowering

Figure 5.4. Unfolding umbel of flowers in common milkweed (*Asclepias syriaca*). For context within the whole shoot, see Figure 5.5. (Photos Craig Holdrege.)

plants—the leaves become smaller in the flowering region. The uppermost leaves contract to a size comparable to the very first leaves on the shoot, but tend to be more pointed and elliptical in shape. These changes in vegetative structure point us to the next developmental wave in milkweed's life history—the flower.

From Flower to Fruit

Among the unfolding upper leaves you can see small grayish-green balls of tightly grouped flower buds (Figure 5. 4). As the buds grow and gradually turn pink in color, the stout stem that carries them away from the main shoot extends diagonally upward. Each flower bud has, in turn, its own delicate stalk, which also lengthens, and the tight ball becomes a looser and larger sphere. All the individual

Figure 5.5. One shoot from a colony of common milkweeds (*Asclepias syriaca*) shown over the course of three weeks during its flowering phase. On June 22 no flowers are open yet and one sees that the lowermost inflorescence on the plant is most developed. After a week (July 1) all the flowers of the lowermost inflorescence are wilting while those in the uppermost inflorescence have yet to open. On July 7 most flowers have wilted. The flower remnants have dropped off the plant or are completely shriveled by July 13; the first developing fruits are visible—only few fruits develop from the approximately 150 flowers in one inflorescence. (Photos Craig Holdrege.)

flower stalks (pedicels) originate at the apex of their common stem, which means that in botanical terms the milkweed inflorescence is an umbel. There can be anywhere between 10 to nearly 200 flowers in one umbel; the average flower number per umbel has been reported as 76 (Willson & Rathcke, 1974) or 104 (Kephart, 1987). While in many umbel-forming plants, as in most members of the carrot family, the flowers spread out into a plane to form a disk, the umbel of the common milkweed maintains its spherical form until the flowers wilt.

As Figure 5.5 shows, the lowermost umbel on a given shoot opens first, then the one next higher up, and so forth. A shoot has an average of 3 to 6 umbels, but some shoots have none and others up to 10 (Hartman, 1977; Willson & Rathcke, 1974). Usually when the lowermost umbel is already wilting, the uppermost one is still in bloom. Milkweed flowers are in comparison to the flowers of other plants long-lived, since any single flower can be open for over a week (Kephart, 1987, gives an average of 9 days; Wyatt & Broyles, 1994, an average of 4 to 8 days). The whole phase of flowering in a colony lasts about 4 weeks.

Long before you come close enough to a colony to be able to see its flowers, you can smell that it is in full bloom. The sweet scent of the nectar-filled flowers carries far. As you approach the colony you see the rich-pink flowering spheres and a multitude of insects crawling and flying around. Honeybees, native bees such as bumblebees, ants, and a variety of butterflies move from flower to flower, umbel to umbel, drinking up nectar out of the blossoms. Along the way they pollinate the flowers, a strangely intricate process related to the complex anatomy of milkweed flowers that I will discuss below.

With the multitude of flowers and pollinators, you would think that many fruits (pods) would develop. But this is not so. Although one shoot may have between 300 and 500 flowers, only few develop into pods—according to Willson and Rathcke (1974), about one pod for every 150 flowers, while Hartman (1977) found that one pod developed for every 60 to 100 flowers. Most flowers in an umbel wilt, while the occasional fruit shows itself through stalk thickening and the white and fuzzy swelling that is the developing pod (Figure 5.6).

Figure. 5.6. The development of the fruit pod in common milkweed (*Asclepias syriaca*). A. The first pods. B. The larger pods lower down are further developed, while the pods from the upper inflorescence are just beginning to form. C. Later in the summer a stem with pods at a variety of stages of development. D. Full-sized pods (about 10cm long); note the upright orientation. (Photos Craig Holdrege.)

Figure. 5.7. Seed release in common milkweed (*Asclepias syriaca*). (Photos Craig Holdrege.)

While vegetative growth is rapid and expansive, and flowering is a period of bursting productivity, pod development is slow and extended. During July, August, and September the pods grow and in their inner cavity the seeds develop. One interesting feature of pod development is that the stalk that carries the pod—regardless of the angle at which it originally extends from the umbel's stalk—twists and curves into a position such that the pod becomes oriented vertically. The pod expands slightly laterally, but mainly grows in length. By October or early November the pods have reached their full size and maturity. The suture along the convex side of the pod splits open and the neatly ordered, tightly packed seeds become visible (Figure 5.7). It looks as if an artist had laid the seeds out. With further opening of the pod, the seeds begin to fall and float away. Each seed has lovely white silky extensions (comas) that allow them to be carried away by a breeze—even though, as seeds go, common milkweed seeds are significantly larger and heavier that those of other old field species (Wilbur, 1976). Interestingly, Kephart (1987) found that although individual shoots within a colony may flower three weeks later than others, the fruits in that colony release their seeds simultaneously.

I mentioned above the paucity of pods in comparison to the wealth of flowers in a milkweed shoot or colony. But each pod is full of many seeds—one study gives an average of 226 (Willson & Rathcke, 1974)—so that in a colony of say 1,000 shoots, hundreds of thousands of seeds will spread into the environment. You would, as a result, expect to easily find seedlings of milkweed plants in areas around existing colonies. As I began to study common milkweed I searched for pictures of seedlings so I could identify them. To my surprise I could find none, although you can find images of the seedlings of most agricultural field weeds. No studies of milkweed I have found describe seedlings in the wild. Evidently, milkweed seeds are not proliferate germinators.

Since, however, new and young colonies can be observed, at least once in a while seeds must germinate and some seedlings take hold. As field experiments show (Agrawal, 2005), already in the second year a plant can produce multiple stems, some over a meter apart from each other, showing the vigorous growth of the rhizome.

In this sketch of its life history we can see how, up through flowering, common milkweed is characterized by exuberant vitality: vigorous underground rhizome growth; yearly expansion of many long, large-leafed shoots; production of numerous flowers that secrete copious amounts of nectar. After flowering, milkweed pulls back and concentrates its vitality into formation of a relatively small number of pods, but each swells into a large pod that houses a multitude of apparently viable seeds that spread into the larger environs. Yet only a few of them form new colonies.

Flower Morphology and Pollination

While the form of the leaves in common milkweed—and other members of the genus as well—is simple, and the shape hardly transforms from bottom to top of the stem, the flowers are highly complex and differentiated. It took much detailed botanical study to identify and relate milkweed flower structure to the anatomy of other flowers (Bookman, 1981; Woodson 1954).

When the flower opens, five small green sepals fold back and then five larger elliptical white to pink petals also fold back and come to lie more or less parallel with the flower stalk (Figure 5.8). In most "typical" flowers of other kinds of plants, you find the circle of petals surrounding a circle of stamens, whose anthers release pollen, and then in the center of the flower the pistil in which the seeds will develop. Neither the stamens nor the pistil are clearly recognizable in the milkweed flower. Rather, a number of accessory organs form out of the partially fused stamens and pistil. Figure 5.9 gives a diagrammatic representation of the milkweed flower.

Above the petals is a so-called corona consisting of five cuplike hoods out of which extend little curved horns. The hoods hold the nectar that attracts so many insects on warm, sunny summer days. In between the hoods are little vertical slits. Each slit opens into a stigmatic chamber. (What it has to do with the stigma, we will see further below.) Above the slit is a tiny black knob, the corpusculum. What one doesn't see is that the corpusculum has two little arms (called translator arms) that extend into the two upper sides of the chamber. Each arm attaches to a golden package of pollen, called a pollinium (pollinia in the plural). Each of the pollinia houses hundreds of pollen grains. Unlike the pollen in most flowers, which is released from the anthers while they are still attached to the flower, in milkweeds the pollen remains contained within the pollinia until

Figure 5.8. A. Flower buds still closed except for one partially open in the lower left hand corner (the same flower as in B). B. two sepals and two petals have folded back. C. An open flower; the petals are not yet fully folded back onto the stalk. D. Side view of a fully opened flower. (Photos Craig Holdrege.)

it comes in contact with the stigma of a flower. Only orchids—also plants with complex flowers—package their pollen in a similar way.

The only way for the pollinia to escape their chambers is via insects. As already mentioned, milkweed flowers are visited by a multitude of insects in search of nectar—which they find in generous supply. Not only is the nectar very rich in sugar content, being up to 3 percent sucrose, but the supply is also renewed over the life of the individual flower (Southwick, 1983; Wyatt, et al., 1992). Milkweed

side view of flower

corona from above

corona from the side with two pollinia visible

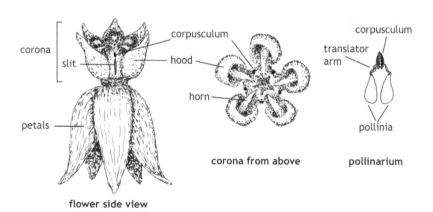

Figure 5.9. Flower structure of common milkweed (*Asclepias syriaca*). See text for description. (Photos and illustration Craig Holdrege.)

flowers produce much more nectar than the many insects feeding on it could ever remove.

While insects are moving around on a flower—they sometimes go from hood to hood on a single flower—one of their legs may slip into a slit. I have often observed a honeybee or a fly struggling to pull a leg out of the slit. Most often it succeeds by pulling its leg upward whereby the leg hooks into a groove on the corpusculum at the top of the slit, and as a result the insect pulls the whole pollinarium—corpusculum, two translator arms and two pollinia—out of the stigmatic chamber. One study (Morse, 1982) showed that a bumble bee picks up on average a new pollinarium every 2 to 5 hours and that the same pollinarium remained attached to the legs for on average 2.5 hours, while if attached to the mouthparts, the pollinarium stayed attached for 10 hours. As an insect continues its nectar foraging, it often accumulates multiple pollinaria. One pair of researchers (Jennersten & Morse, 1991) reported finding up to 35 pollinaria on a single insect! Sometimes they are hooked together in chains of ten or more dangling from an insect's leg. Most often one sees honeybees with many pollinaria attached to them (Figure 5.10).

Figure 5.10. A. A honeybee moves through an umbel of milkweed flowers. B. Close-up of a honeybee on a milkweed flower with multiple pollinia dangling from its legs. (Note the slit and corpusculum (dark spot) between two of the hoods of the flower in the lower left of photo B). (Photos Craig Holdrege.)

Smaller insects—I have observed mainly flies—may be unsuccessful in removing a leg from a slit and the limb will tear off in the process. I have also seen dead flies hanging from a flower with a leg still caught in a slit. Larger pollinators like bumblebees and honeybees are more likely to extract pollinaria without losing a body part.

What happens to the pollinaria? Many will simply drop away or be rubbed off as the insect moves around. A few will find their destination in another flower. And just as the removal of the pollinaria is both a haphazard and narrowly constrained process, so also is pollination. First, something remarkable and absolutely essential for pollination occurs with the pollinarium itself. After an insect has been carrying around a pollinarium for about ninety seconds, the pollinarium dries out, and in the process the translator arms rotate ninety degrees (Wyatt & Broyles, 1994). This torsion is highly significant because it brings the pollinium into a position that allows it to slip into a slit when an insect moves over one. Before the torsion, the broad side of the pollinium usually faces the slit as the insect crawls along; afterward the narrow side of the pollinium is in line with the slit so that if the insect is moving at just the right height over a slit, the pollinium can slide into the stigmatic chamber. As the insect moves ahead, the translator arm breaks and the pollinium remains behind. Once in the chamber, one edge of the pollinium rests against the receptive wall of the stigmatic chamber. While in most flowering plants the stigma is open to direct contact with the air and visiting insects, the five receptive stigma surfaces in milkweeds are enclosed within five stigmatic chambers, open to the world only through the narrow slit. So in milkweed flowers, both pollen grains and stigmatic surfaces are housed in enclosed structures.

Only when the pollinium slides into the stigmatic chamber can this encapsulation be overcome. Everything I will now describe is visible only when one dissects flowers at different stages of the pollination process and examines the structures under a microscope (Sage & Williams, 1995). The stigmatic chamber is the source of nectar for the hoods, and when the pollinium is inserted into the chamber it is bathed in nectar and begins to swell. Within a few hours the edge in contact with the receptive inner surface of the stigmatic chamber breaks open and multiple pollen tubes grow out of the pollen grains. The tubes grow down the style and into one of the two ovaries of a flower. An ovary contains a couple hundred ovules (Baum, 1948) and each one needs to be fertilized by the nucleus from one pollen tube. Interestingly, when fertilization occurs, virtually all of the ovules are fertilized—one rarely finds milkweed pods with just a few seeds (Wyatt & Broyles, 1994).

One could at first think that the highly specialized pollination apparatus, which provides a seemingly perfect fit between pollinium and stigma, would guarantee successful fruit set. But this is not the case. In the first place, the specialization also means that relatively few pollinia actually are successfully inserted into a stigmatic chamber. Secondly, researchers discovered that common milkweed is self-incompatible (reviewed in Wyatt & Broyles, 1994), meaning that the pollinium inserted into a flower of the same colony will normally not bear fruit. The pollen tubes grow into the ovary, but the seeds do not develop. Because insects are moving largely within a given colony, it is likely that most of the pollinia inserted come from flowers of the same colony and will therefore not lead to successful fruit set. One study (Pleasants, 1991) using radioactively labeled pollinia found that a third of the pollinia inserted were from the same umbel and most of the other inserted pollinia came from the same colony; only a few were carried to other colonies and successfully inserted.

If this "inefficiency"—we could also call it overabundance—were not odd enough, even when researchers cross-pollinated flowers of related *Asclepias* species by hand, fruit set never exceeded 20 percent and was usually considerably lower (Wyatt & Broyles, 1994). Thus, overall, the specialization of milkweed pollination, at least in relation to successful fruit formation, is connected with a low vitality. If one looks, however, at a milkweed colony from the perspective of the many insects that feed on the flowers' nectar, milkweed is contributing significantly to the vitality of those animals. Of course, the picture is more complex if we think of the occasional death of mainly smaller insects unable to remove a limb caught in a slit.

One interesting feature of the relation between flower structure and insects is that, although all milkweeds have such specialized flowers, they are visited by and can be pollinated by a wide array of insects. The nectar also attracts ants and small beetles that usually contribute little toward pollination, but reap the benefits of the ample nectar supply (Fritz & Morse, 1981).

Abundant Animal Life

We have already seen that common milkweed is an important part of the life of insects that feed on its nectar. Observing nectar feeders on common milkweed, Southwick (1983) identified representatives from 15 different orders of insects (and one hummingbird species). Nectar was taken mainly during the day but also during the night by a variety of nocturnal moths. But these nectar-feeding insects represent only a minority of the insects and other arthropods that interface with milkweed. In the late 1970s Dailey and his colleagues carried out surveys of bugs (Hemiptera) and beetles (Coleoptera) found on common milkweed (Dailey, Graves, & Herring, 1978; Dailey, Graves, & Kingsolver, 1978). Over the course of ninety days, they found 132 different species of beetles, 18 of which they considered common visitors, since they collected more than 50 specimens of each. They collected 45 species of bugs, 5 of which were common visitors according to the same criterion. Milkweed teems with insect life.

For many insects, milkweed is certainly a small and transient part of their habitat—or speaking functionally, a minor part of their ecological niche. They may nibble on the leaves and flower buds or drink some nectar and then move on to other plants. As predators they may, like the bug *Phymata fasciata*, hide in the thicket of milkweed stems, leaves, and flowers, waiting for their prey of flies and small wild bees. And then there are the milkweed specialists, which I will discuss below, that feed almost exclusively on milkweeds. So milkweed provides food and a microhabitat for a multitude of organisms. Its exuberant growth—in rhizomes, stems, leaves, flowers, fruits, and seeds—allows abundant insect life to orient around it.

The Extended Organism

One striking feature of the common milkweed is that the entire life cycle of a number of insect species is tightly interwoven with it. There are at least 10 species of insects that feed only on common milkweed or other closely related milkweeds in the genus *Asclepias* (Agrawal, 2005; Price & Wilson, 1979; see Table 5.1 and Figure 5.11). None of these specialist species is a nectar feeder; rather, they feed on milkweed rhizomes, shoots, leaves, flowers, or seeds. The most well-known of these is the monarch butterfly (*Danaus plexippus*). The adult butterfly lays its eggs on the leaves of common milkweed, the larvae live from its leaves and the milky sap the plants contain, and the adults drink from the flower nectar, although adults are not restricted to milkweeds.

What is fascinating about the monarch and some of the other milkweed specialists is that they do not just feed on the plants, digest the substances, and then build up their own body substances. Rather, they store some of the components of the milkweed sap in their body. When a milkweed stem or leaf is damaged, it exudes a white sap. All you have to do is to scratch the stem with your fingernail and the white sap oozes out and streams down the stem until it gradually hardens. When, for example, a monarch larva bites into a leaf vein or stalk, the sticky (latex-containing) milky sap seeps out and the larva ingests it. It draws out of the sap a particular group of substances known as cardiac glycosides (cardenolides), and instead of breaking them down or excreting them, it stores them in its tissues. The concentration of cardiac glycosides in the tissues of a monarch is substantially higher than it is in the tissues of common milkweed. Interestingly, it is not only the larva that sequesters these substances; they are also retained in the adult, which has gone through the complete metamorphosis from caterpillar to butterfly. So part of the milkweed becomes an essential part of its insect predators.

Caterpillar (left) and adult (right) of the monarch butterfly (*Danaus plexippus*)

Red milkweed beetle (*Tetraopes tetrophthalmus*)   Milkweed leaf beetle (*Labidomera clivicollis*)

Nymph (left) and adult (right) of the large milkweed bug (*Oncopeltus fasciatus*)   Milkweed tussock moth larvae (*Euchaetes egle*)

Figure. 5.11. Different insect specialists associated with common milkweed. Not presented at the same scale. (Photos Craig Holdrege.)

Table 5.1. Milkweed-specific herbivores

Species	Primary Larval Food	Conspicuous? (Aposematic; larva & adult)	Sequesters cardiac glycosides?	References
Monarch Butterfly (*Danaus plexippus*)	Foliage of common milkweed and other milkweed species	Yes	Yes	Malcom, 1995
Milkweed Tussock Moth (*Euchaetes egle*)	Foliage of milkweeds and dogbanes (*Apocynum*)	Larva, yes; Cryptic adult	Larva: yes; Adult: little	Barber & Conner, 2007
Milkweed Tussock Moth (*Euchaetes egle*)	Foliage of milkweeds and dogbanes	No (larvae beige; adult white)	Yes	Cohen & Brower, 1983
Red Milkweed Beetle (*Tetraopes tetrophthalmus*)	Larva: rhizomes of common milkweed; adult: leaves and flowers of common milkweed	Yes	Yes	Hartman, 1977; Farrell & Mitter, 1998
Milkweed Leaf Beetle (*Labidomera clivicollis*)	Foliage of common and swamp milkweeds	Yes	No	Berenbaum, 1993
Seed Weevil (*Rhyssomatus lineaticollis*)	Pith common milkweed stems	No	No	Fordyce & Malcom, 2000
Large Milkweed Bug (*Oncopeltus fasciatus*)	Seeds of common and other milkweeds	Yes	Yes	Duffey, et al., 1978
Small Milkweed Bug (*Lygaeus kalmii*)	Sap of common milkweed (not a narrow specialist)	Yes	Yes	Fox & Caldwell, 1994; Wilbur, 1976; Duffey & Scudder, 1972
Milkweed Aphid (*Aphis asclepiadis*)	Inside of milkweed leaf tissue	No (cryptic)	Unknown	Mooney, Jones, & Agrawal, 2008
Leaf Mining Fly (*Liriomyza asclepiadis*)	Inside of milkweed leaf tissue	No (cryptic)	Unknown	Agrawal, 2005

Cardiac glycosides are bitter tasting and can disrupt the ionic balance of a number of different cell types in animals, including heart muscle, vascular smooth muscle, neurons, and kidney tubules (Malcolm, 1995). In high doses they can be fatal to an animal, but in nature this will rarely happen, since they cause vomiting in prelethal doses. We would imagine that common milkweed is protected against herbivores by the cardiac glycosides in its sap. Clearly, however, the sap does not prevent specialist herbivores from feeding on milkweed and sequestering cardiac glycosides, although some of these specialists avoid taking in large amounts of sap while feeding. The monarch and red milkweed beetle are known to bite into a milkweed leaf vein near the base of the leaf, which then exudes sap that flows back out of the more distally located veins (Dussourd, 1999; Helmus & Dussourd, 2005). The insect then crawls to the periphery of the leaf and begins to feed from the part of the leaf that now contains little sap.

Unsurprisingly, researchers believed that by sequestering cardiac glycosides, milkweed predators may be protected against their own predators. Beginning in the 1960s researchers began testing this hypothesis and, as Malcolm concludes in a review, "Much evidence is published to show that many prey species are well defended against predators by the presence of cardenolides" (1995, p. 101).

So milkweed is helping those insects that prey on it become better protected from their own predators. This is, in a sense, a paradoxical situation in which a plant is providing protection for its predators, which increases the likelihood that there will be more predators to feed on it (Malcolm, 1995). Theoretically, one could think that these specialists might eradicate milkweed. But neither the scientific literature nor my own observations indicate that milkweed populations are significantly harmed by the specialist herbivores associated with them. And it is not as if the monarch or other milkweed specialists have no predators. Prysby (2004) points out that both monarch adults and larvae are preyed upon at least occasionally by some birds, mice, ants, dragonflies, and wasps, and that the larvae can be parasitized by flies and wasps.

Most of the milkweed specialists that sequester cardiac glycosides are brightly colored (Figure 5.11). (Within a Darwinian framework, one interprets such coloring as warning coloration, also called aposematic coloration. The theory is that the bright colors and patterns evolved as a warning sign to predators that signals "keep off.") Hartman (1977) noticed an additional correlation, namely that the brightly colored, cardiac glycoside–storing herbivores tend to move around the plant a good deal when feeding, eating only small amounts and rarely doing significant damage even to a single shoot. The conspicuous caterpillars of the milkweed tussock moth, in contrast, aggregate on a shoot and can denude it of leaves, leaving only the skeleton of the larger veins. Interestingly, tussock moth caterpillars, which sequester cardiac glycosides, metamorphose into inconspicuous (cryptic) nocturnal moths that do not sequester appreciable amounts of cardiac glycosides (Barber & Conner, 2007).

As an adult, the monarch butterfly migrates south. The monarchs east of the Mississippi fly as far as 4,800 km to Mexico, where they overwinter. "Amazingly, these butterflies fly from their summer breeding range, which spans more than 100 million ha [hectares], to winter roosts that cover less than 20 ha, often to the exact same trees, year after year" (Solensky, 2004, p. 79). The expansive extent of the summer range corresponds to the range of common milkweed and a number of other milkweed species. Along the route of their migration, they feed on milkweed nectar and the nectar of other flowers. Their range contracts to the small overwintering area in Mexico, where they are temporally and spatially separated from milkweed. However, they still carry small traces of the plant in their bodies through the cardiac glycosides. The next spring they migrate back north and many of these adults mate, lay eggs, and die in the southeastern U.S. Their offspring feed on southern milkweeds, metamorphose, and the adults fly north to find common milkweed flowering in the northern summer. The life cycle begins anew.

While the life history of an individual monarch can span nearly a whole continent, the life history of a red milkweed beetle (*Tetraopes tetrophthalmus*) is much more tightly linked to a local common milkweed population (Agrawal, 2005; Farrell & Mitter, 1998; Hartman, 1977). About the time a colony of milkweeds begins to flower, bright-red milkweed beetles crawl out of the ground and spread out onto milkweed shoots—an insect version of flowering. They crawl around on the plants and may fly short distances. They generally don't leave the area of the colony. They begin feeding—they feed some on leaves (Figure 5.11), but mostly they feed on flowers. When a milkweed colony is at a high point in flowering, the red milkweed beetle has its peak in population density. The adults live for about three to four weeks, which corresponds to the main phase of flowering. The synchrony between adult beetle and flowering milkweed is striking. In a colony that flowers later in the year, the beetles emerge later. It could be that the temperature of the soil helps to coordinate this synchrony, since both shoot development in milkweeds and pupation in the milkweed beetle are temperature dependent (Hartman, 1977).

The beetles mate and the female moves to a nearby grass plant or other hollow-stemmed old field plant and nibbles a hole in the stem, crawls inside, and lays her eggs. This is the one phase of the life cycle that is not dependent on milkweeds. When the eggs hatch, the larvae crawl down into the ground and move to the milkweed rhizomes. There they begin to feed, on both the inside and outside of the rhizomes. They feed exclusively on milkweed rhizomes. They can do considerable damage to short sections of a rhizome, but never have a significantly detrimental effect on a colony as a whole. While the colorless larvae are busily feeding below ground on the rhizomes, the fiery red adults have died. The larvae feed until early fall, when they move out of the rhizomes and overwinter in the soil, near the rhizomes, as large pre-pupae. They do not feed during this time. Both milkweed and pre-pupae are quiescent during the winter. Only when the soil reaches a temperature of around 63 degrees Fahrenheit (17 to 18 degrees Celsius) does the pre-pupa become active—not through movement or feeding, but through metamorphosis. It forms a pupa out of which the adult beetle soon emerges. It breaks through the cocoon and digs its way out of the soil to emerge in a forest of milkweeds, where it begins to feed. The next adult generation begins its short life.

When we reflect on such relationships between two kinds of organisms, a plant and an animal, the boundaries between the two begin to dissolve. We can no longer think of the plant without the animal and the animal without the plant. Normally we think of the plant and the animal that feeds on it as two separate organisms that interact. It is challenging, in fact, not to describe them in such terms. But we can ask the question, "Where do organisms end?" (Holdrege, 2000a). Clearly, the milkweed is unthinkable without its animal associations, just as the animals cannot be described or understood without the milkweed. Milkweed's pollination is wholly dependent upon insects, just as many insects are dependent upon milkweed for food and reproduction. Therefore, we must transcend the boundaries we construct when we look at an organism from a taxonomic standpoint. We can begin to see organisms as intersecting relationships that are part of the greater web of life. In the case of common milkweed, these relationships are especially evident, since even some of its physical substances (cardiac glycosides) become a part of various animal species.

From an evolutionary perspective, we need to imagine that the lives of common milkweed and its specialist insects have been related to each other for a long period of time—going back to the mid-Tertiary in the case of the red milkweed beetle (Farrell & Mitter, 1998). They have coevolved and have a history together—they belong to each other or are part of each other. One of the key realizations of an ecological-evolutionary perspective is that what appear today as separate entities are in fact interconnected. As one biologist has stated, "The process of co-evolution between plants and their natural enemies—including viruses, fungi, bacteria, nematodes, insects and mammals—is believed by many biologists to have generated much of the Earth's biological diversity" (Rausher, 2001, p. 857). That this diversity is an expression of the interconnectedness between life forms is what we begin to understand and to appreciate when we concern ourselves with the life histories of intersecting organisms.

Summarizing Picture

When you see an old field, the robust common milkweed plants stand out among the much sleeker grasses, asters, or goldenrods. Common milkweed has thick stems and expansive leaves that in shape and size look more like the leaves of a plant growing in shady woods than one growing in a sunny old field. In the warm summer days of late June and through much of July, the large spherical heads of flowers unfold on the upper part of the stems. The individual flowers are actually quite large for a field plant and they produce large amounts of concentrated nectar. Their scent spreads out into the surroundings. When in flower, a colony of milkweeds attracts—day and night—a great variety and number of insects of all different shapes and sizes. For several weeks in summer milkweed becomes a microhabitat with a singular concentration of insect life.

The flower is highly specialized. Those parts of the flower that normally are in direct contact with the air and insects—the receptive stigma and the pollen grains—are encapsulated, the stigma within the stigmatic chamber that opens to the world only through a narrow slit and the pollen grains in the pollinia, which themselves are hidden within the chambers. Pollination becomes an intricate process of removal and insertion that is unthinkable without the intervention of insects. Only they can bring the specialized structures into the precise spatial relation the plant needs for fertilization to occur.

While the flower outwardly displays milkweed's strong specialization in its form, all parts of the plant except the flowers produce the specialized latex sap. While the flowers produce no latex sap (S. Malcolm, personal communication), they do produce the sugary nectar. The latex sap is encountered by animals that feed on the plant. Small insects can become caught in the sticky sap. Others can be repelled by the cardiac glycosides in the sap, while still others incorporate the toxins into their own bodies. The life of these mostly vibrantly colored insects is in multiple ways closely bound up with the milkweed.

After the flowers wilt, the fruit pods begin to expand. While relatively few fruits form out of the multitude of flowers, those that develop grow large—much larger than those of other old field-community plants. The pods swell and orient themselves upward, a contrasting gesture to the hanging inflorescences. Each pod is full of seeds, seeds that are large and heavy. But they have the light feather-like extensions of the white comas that allow them to be carried away on a breeze when the pods split open. It is almost as if the upward pointing pods are prefiguring what is to come—the upward lift of the seed-bearing comas that disperse the seeds into the larger environment. As with all stages of milkweed, both pods and seeds provide nourishment to insects.

One salient feature that informs milkweed is its exuberant and robust growth. Underground it spreads year to year, forming a network of thick rhizomes out of which the aboveground shoots grow. The thick shoots bring forth large, spreading leaves. All these parts of the plant contain the milky sap, which is continuously produced as the plant grows and develops. A marked transformation in substance and form occurs as the many large umbels unfold in the summer light and warmth. Just as the stems and leaves are rich in milk sap, so are the flowers rich in sweet nectar. Both the leaves and the flowers attract countless insects; milkweed is of fundamental importance to the existence of some of these creatures. In the fall, large pods form, containing many large seeds that spread out into the environment.

Milkweed is effusive and yet it is also specialized. This specialization both attracts and repels insects. Think of the sticky, toxic sap that can also be protective, or the pollination process in which insects are attracted to the nectar, but may become injured or trapped by the flower structure. Milkweed invites life, but also holds it back. There is a fascinating tension in this plant.

Leopold's "Thinking Like a Mountain"

I began this chapter with a quotation from Aldo Leopold:

All I am saying is that there is also drama in every bush, if you can see it. When enough men know this, we need fear no indifference to the welfare of bushes, or birds, or soil, or trees. We shall then have no need of the word "conservation," for we shall have the thing itself. (1999, p. 172)

Leopold could make this bold statement because he had experienced in a clear and deep way the "drama" in every organism—how each organism is a whole that is active within the larger living context of its environment. I have tried to portray something of the drama of the common milkweed. Such a whole-organism study leads into both depth and breadth. We get nearer to the specific qualities of the plant. We begin to see its uniqueness within all the details and begin to articulate those unique qualities. Inasmuch as we are able to do just that, a story of this organism's unique way of being emerges—no other plant is the same. Becoming aware of such a story and participating in it cannot leave us cold—we have met a unique quality in the world and the world would be poorer without it. Holistic knowing creates the basis of a moral relation to the world, for when we have experienced nature in this way, "we need fear no indifference to the welfare of bushes, or birds, or soil, or trees."

The story of an organism always leads beyond itself to a larger web of relations with other organisms and elements of the environment. There is no isolation in the living world. We attend closely to the specific qualities, for instance, of milkweed, monarch butterfly, or milkweed beetle, and at the same time become vividly aware of how these qualities intersect and are mutually dependent. We begin to gain insight into the truly ecological nature of life.

Perhaps the most succinct and vivid portrayal of the deeply ecological nature of life I know is in Leopold's famous essay, "Thinking Like a Mountain." (Leopold wrote this essay in 1944; it was still unpublished when he died in 1948 and then included in the first edition of *A Sand County Almanac*, which was published in 1949.) In this essay Leopold portrays the wolf and wolf country:

Only the mountain has lived long enough to listen objectively to the howl of a wolf. Those unable to decipher the hidden meaning know nevertheless that it is there, for it is felt in all wolf country and distinguishes that country from all other land. It tingles in the spine of all who hear wolves by night, or who can scan their tracks by day. Even without sight or sound of wolf, it is implicit in a hundred small events: the midnight whinny of a pack horse, the rattle of rolling rocks, the bound of a fleeing deer, the way shadows lie under the spruces. Only the ineducable tyro can fail to sense the presence or absence of wolves, or the fact that mountains have a secret opinion about them. (1989, p. 129)

For Leopold the wolf is not a separate organism that outwardly interacts with other organisms and the landscape. The wolf is *present* (or is a presence) in the whole landscape. The wolf participates in all aspects of what Leopold calls the mountain—he is thinking of a rugged landscape in eastern Arizona. The other organisms—the deer, the horses, the humans—express "wolf" in their activity and sensations. And the rest of the landscape—the rolling rocks, the shade beneath the trees, the tracks in the ground—also expresses "wolf." Leopold has experienced the landscape as a living, interwoven whole.

In calling his essay "Thinking Like a Mountain," Leopold is pointing to a quality of thought that leaves object thinking behind it. Yes, it is important—and was important for Leopold—to understand intellectually predator-prey dependencies and how they affect the long-term character of an ecosystem. But this is not yet "thinking like a mountain," because it is still bound to intellectual categories. Thinking like a mountain demands a participatory way of knowing in which the human senses and human mind can transcend a spectator awareness and actually partake in the intersecting and vital qualities that are part of every landscape. Thinking like a mountain is a thinking alive with the world. Thinking like a mountain is neither an abstract idea nor a poetic-sounding metaphor. It is, I suggest, at its core the same kind of concrete perceiving/thinking that allowed Goethe to see the archetypal plant—the fuller expression of plantness—in each of the many different plants he encountered in his studies and travels. The world as connectedness reveals itself.

Leopold's initial striving to think like a mountain was sparked by an experience he had with wolves when he was twenty-two years old (in 1909; Meine, 1988, chapter 6). He was a new crew chief for the forest service in the Arizona Territory (later the state of Arizona). He and his crew were responsible for carrying out an inventory of the forests and timber in the rugged mountains of eastern Arizona. In "Thinking Like a Mountain," he describes being with one of his crewmen when they saw what they thought was a deer crossing through the turbulent whitewater of a river. When it reached the bank, they realized they were watching a wolf, which was greeted by a half dozen grown pups. "In those days we had never heard of passing up a chance to kill a wolf. In a second we were pumping lead into the pack" (Leopold, 1989, p. 130). They hit the adult wolf and went down to it:

> We reached the old wolf in time to watch a fierce green fire dying in her eyes. I realized then, and have known ever since, that there was something new to me in those eyes—something known only to her and to the mountain…. I was young then, and full of trigger-itch; I thought that because fewer wolves meant more deer, that no wolves would mean hunters' paradise. But after seeing the green fire die, I sensed that neither the wolf nor the mountain agreed with such a view. (p. 130)

In this face-to-face experience with the wolf, Leopold's preconception of the wolf as a harmful predator for a moment dropped away. He was open to the wolf, and something spoke through "the fierce green fire dying in her eyes." This was much more than a "mere" sense impression. A being communicated itself to him, or, we could also say, in that moment he was participating in the wolf and through the wolf in the mountain. And this encounter, this interaction, touched the depths of his soul. He "sensed" that the mountain and the wolf did not agree with his view. He transcended the utilitarian, human-centered thought construct that brought him to meet the wolf in the first place—the idea of removing a predator to have more deer to hunt. He gained a new awareness of "wolf" as something much more than a roaming predator.

The deep shift that occurred in Leopold on this day—or perhaps we could say, the day on which the mountain and wolf became part of Leopold—was an experience that initiated a long process of development and with time bore much fruit. He was not transformed overnight, and it is not as though all his subsequent thinking simply followed from this experience. But it was a key event that brought about a shift. This shift led to a gradual transformation in his view of wolves and was at the core of his overall ecological view of life, which continued to evolve over decades. Leopold came to see how humanity is interwoven into the totality of nature. His thinking culminated in what he called a "land ethic," an attempt to formulate the basis of an ethical relation of humanity to the rest of the biotic community. He writes: "A thing is right when it tends to preserve the integrity, stability, and beauty of the biotic community. It is wrong when it tends otherwise" (Leopold, 1989, pp. 224–225). This simple, forceful statement—written in 1949—characterizes sustainability long before the word existed. It only makes sense when we have glimpsed to some degree what Leopold means by "thinking like a mountain," thereby giving substance to ideas like "integrity" or "beauty," which otherwise remain abstract. Such concepts must be rooted in full-blooded, concrete experience. In a well-meaning attempt to bring Leopold's formulation up-to-date, Callicott rephrases the land ethic in the following way: "A thing is right when it tends to disturb the biotic community only at normal spatial and temporal scales. It is wrong when it tends otherwise" (2002, p. 104). In this reformulation all the freshness, vigor, and force of Leopold's statement have disappeared; it has died into an academic abstraction.

Leopold writes that "no important change in ethics was ever accomplished without an internal change in our intellectual emphasis, loyalties, affections, and convictions" (pp. 209–10). This makes clear that a "sustainability ethic," as we might call it today, demands such holistic transformation in the human being—it has its wellspring in a living, participatory way of knowing.

As Leopold's example of thinking like a mountain shows, truly transformational experience is experience that takes hold of us in the present and engages our senses, our body, our thinking, and our emotions. As a whole-person experience, it is a uniting of the outer and the inner that overcomes dualism: in the experience itself there is no inside and outside; it is "me-in-the-world," the "world-in-me," or, even better, "world-me." This is the lived overcoming of the object view of the world and affirmation of the participatory nature of reality.

Transformational experience works beyond the moment into the future. It stays with us, not as a new content in an old container, but as a broadening and deepening of ourselves through the world. And to the degree that we are touched and changed by the world in experience, our ability to engage in ever new transformation through new experience grows.

CHAPTER 6

Conclusion: A Quiet Revolution

The Plant's Teachings—A Summary

THE STUDY OF PLANTS in this book has led us into a realm of forms, processes and transformation, relational dynamics, and unique beings. When we notice, carefully observe, and actively follow the way plants live, they become part of us. We learn not only about them but from them. In this sense plants become our teachers. We meet life in the plant and this life can light up in us and increasingly inform the way we interact with the world. This is what I call living thinking. In developing living thinking we begin to participate in the life of the world and the mental barriers that separate us from it and lead us to view the world in abstract, static, and decontextualized ways gradually dissipate.

The plant lives in intimate connection with its environment and is sustained by this intimate relation. We become rooted in the world through careful and open-minded exploration and experience of the sensory world. We turn toward the concrete appearances and happenings of the world. We remain open to and orient ourselves around—stay in touch with—the phenomena with which we are interacting. The active and ever-renewed immersion in sense experience nourishes our inner growth, keeps us vibrant and fresh, and helps us to overcome what has become all-too rigid in our consciousness. By engaging in the concrete, we can escape the grasp of the abstract.

When we attend to plants out of this orientation of mind, they reveal to us life-as-transformation. Our thinking comes into movement, and we learn to form and re-form our ideas. When plants help us become attuned to process, we gain a capacity to discover in any field of inquiry a wealth of dynamic and transformative processes that static and additive ways of thinking would never see. Process thinking is not murky. Just as the plant forms distinct parts but lets them go again, we form clear and distinct ideas as phases in a process, but we can also let them wilt and die away as the plant of living knowledge develops.

The plant is a rhythmical being. It forms leaf after leaf, and it alternates between expansion and contraction in its development. In living thinking we become aware of how a rhythmical process of inquiry helps to enhance learning. When we carefully observe phenomena we expand out into them, we live with them, and let them become part of us. Then we draw back and work with what we have taken in. This prepares us to go out to the phenomena again. I have described exact sensorial imagination as a way of intensifying our experience of the phenomena. We form a vivid picture of, say, the way a plant grows. This connects us more deeply with what we have observed. We also reflect on our experiences, we struggle to understand, riddles arise, we read what others have written, we speak with colleagues. In these ways we come into inner movement that intensifies our receptivity to new aspects of the phenomena when we return to observation.

As we learn through the ongoing rhythmical inquiry into things, the plant of understanding grows. For a long time we may be spreading our roots, growing leaf after leaf, and also shedding leaves in getting to know an area of life. When the plant forms flowers, it enters a new phase of its development. In the flower, leaf-like parts (sepal, petal, stamen, pistil) appear highly transformed as a new patterned whole. In the process of knowing, moments of insight can come— "aha!" experiences—in which the different aspects of what we have been studying appear in a new unity. These moments of insight are the flowers of understanding. Their beauty can enthrall us, but they can do more. When we say that moments of insight and understanding bear fruit, we are pointing to their generative nature, to the seeds of new life they bear. They are a source of inspiration and give direction to new pathways of inquiry and action.

A plant does not simply unfold its forms; it forms itself through its environment. A plant does not exist without an environment. It brings this context to expression in its own form and substance. The plant can adapt to myriad conditions, in each instance bringing forth different aspects of its own plastic nature as well as expressing the unique circumstances of its environment. Likewise, living thinking is in intimate conversation with the phenomenal world. It is truly context sensitive. We are open to what the phenomena have to say; they enrich us, and we, in turn, bring about changes in the contexts we live in and interact with. Each person (and one person at different periods of life) brings something unique into the vibrant conversation. Different aspects of the one world can reveal themselves to each person, just as different species of plants bring the environment to expression in a variety of ways.

The careful study of plants leads us to meetings with remarkable beings, such as milkweed. Every plant we carefully attend to can reveal itself to us as a unique way of being in the world. And as a unique being it is not a separate thing. It is part of a larger world, and its life intersects with the life of myriad other creatures—think of milkweed's relations with insects. Once we begin to see such relations, it becomes clear that the world is a dynamic flux of intersecting beings, and by "being" I mean a configured and self-configuring source of activity.

By growing into all of these relations, our sense of connectedness with the world intensifies. We lose distance and gain a vivid knowledge of the intricate world of which we are a part. This becomes the basis for an ethical relation to an "other." On the one hand, we experience the other's nature as something in its own right; therefore we know it as an "other." But, on the other hand, we know it precisely through connecting with it; we overcome separation, and an ethical fabric forms—a fabric of concern and respect—in the world.

Inasmuch as we enter into such engagement, we realize that life places real demands on our ability to fathom dynamic processes and relations. In other words, a plant we attend to is essentially saying to us: if you want to participate in and become aware of the way I live in the world, you are going to have to change. This awareness is not only a healthy antidote to human arrogance, but can also stimulate us to develop transformative practices of the kind I have described in this book, such as observation exercises or the practice of exact sensorial imagination. We find ourselves on a pathway to develop organs of perception in ourselves so that we can do justice to the world in our perceiving, thinking, and acting.

A crucial part of this transformational work is to cultivate awareness for the mental activity that informs all our interactions with the world. We cannot be truly alive in thought if we are aware only of thought products (concepts, models, etc.). Just as we begin to see the plant as a dynamic process, we can learn to see our mental life as unfolding, highly contextual activity. It is through thought becoming aware of itself that the nature of our embeddedness in the world becomes transparent. It allows us to realize that we are beings amongst fellow beings.

All this is part of the quiet revolution that plants stimulate in us. It is a revolution because everything changes when the object view of the world falls away and an organic view emerges. We are not just adding some new insights onto existing ones. As a quiet revolution, it is not flashy and will not make headlines by propounding new universal theories or programmatic solutions to global problems. But this does not negate its potency. For a fundamental shift in the way individuals relate to the world means sowing the seeds of new life into culture. When the quiet revolution begins, we notice the vacuity of much-cherished abstractions and free ourselves to gain a fresh view of the world that then appears as a fully participated reality in its dynamism and contextuality.

Clearly, much in our times works against this quiet revolution, and the most fertile soil for the growth of living thinking in our culture would be a transformation in the way we view and practice education. While that is the topic of another whole book, I would like nonetheless to conclude this book by characterizing in broad strokes some perspectives on education that emerge out of what I have considered. I will also draw on my twenty-one years of experience as a high school teacher in Waldorf schools, my work in teacher education, and all those joyful and challenging experiences involved in being a father of three (now grown) children. So my question is, what characteristics make themselves apparent when, in the mode of living thinking, we turn our attention to education?

Education and the "Treacherous Idea" of Preparation

In living thinking we develop a kind of sensorium for prevailing abstractions that provide all-too narrow and rigid frameworks for the way we view things. One of those abstractions in education is the guiding notion of preparation. U.S. President Barack Obama's education webpage offers a clear message about the goals of education:

> A world-class education is the single most important factor in determining not just whether our kids can compete for the best jobs but whether America can out-compete countries around the world. America's business leaders understand that when it comes to education, we need to up our game…. The President will reform America's public schools to deliver a 21st Century education that will prepare all children for success in the new global workplace. President Obama's [plan] fosters critical thinking, problem solving, and the innovative use of knowledge to prepare students for college and career, helping America win the future by out-educating our competitors. (http://www.whitehouse.gov/issues/education; downloaded May 3, 2012)

Here the goals of education are framed solely in terms of preparing students to achieve economic success and to serve national interests. Good education, in this view, will generate citizens who serve the economic engine that drives the United States in its efforts to out-compete the rest of the world. This is a crass perspective, but it also indicates a pulse of our times when educational policies focus increasingly on specific outcomes.

In such a frame of mind, thinking about education becomes narrow. Each stage of the educational process becomes the preparation for the next: kindergarten prepares for elementary school, which prepares for middle school, which prepares for high school, which prepares for college, which prepares for a profession. When curricula are developed from this perspective, the tendency is to bring what is perceived as needed at a later stage into an earlier one. A public-school teacher in the United States may now receive training to teach her students how to use PowerPoint in the second grade! Why? Well, they will do their middle school reports using PowerPoint, so they need to be prepared. And why should they do PowerPoint in middle school? They need it for high school ...

Or, in public high schools, there are advanced placement courses so that the students are better prepared for college and can even skip some college courses. In reality, it is often the case that students nonetheless learn the same subject matter again in college courses. Or even worse, as a university chemistry professor once told me: I need to help students who have taken advanced placement courses unlearn what they think they know so that they can actually learn to think like chemists!

When education is mainly viewed as preparation for a next stage of education, for a particular professional outcome, or for furthering national interests, then the student is viewed as an entity to mold to fit a particular system. The goal, as a guiding picture, is given (although rarely filled out and concretely imagined), and the educational program is to lead to this goal. This way of thinking dominates in Western-style education and also provides the prevailing model for schooling worldwide. It makes the future—as the goal to be reached—into something specific and bounded that we can have a grip on. We could call this the abstract future. At the same time, educational experience in the present degenerates into preparation for this abstract future. The "now" is always for a surmised later. John Dewey, who was keenly sensitive to education as a living process, called the notion of preparation a "treacherous idea":

> When preparation is made the controlling end, then the potentialities of the present are sacrificed to a suppositious future. When this happens, the actual preparation for the future is missed or distorted. The ideal of using the present simply to get ready for the future contradicts itself. It omits, and even shuts out, the very conditions by which a person can be prepared for his future. We always live at the time we live and not at some other time, and only by extracting at each present time the full meaning of each

present experience are we prepared for doing the same in the future. This is the only preparation which in the long run amounts to anything. (Dewey, 1997, p. 49; written in 1938)

In other words, real learning occurs through people engaging in experiences that are meaningful in and of themselves. In this way growth occurs and we can interact more fully—be present and creatively engaged—with what comes to meet us in new situations in the future.

Learning from Genuine Presences

When we shed the abstract idea of preparation, a new field of questions opens up. Let me start with one: what kinds of experiences are especially potent in their ability to allow learning and growth to occur? To ponder this question is critical, especially in our times, which—at least seen superficially—offer a plethora of potential experiences. In my view, the best education will bring learners into interaction with what philosopher Albert Borgmann characterizes as

'reality' taken in the sense of genuineness, seriousness, or commanding presence, the sense we have in mind when we speak of real gold as opposed to things that merely glitter and of a real person, a mensch, as opposed to a dude.... What is eminently real has a commanding presence and a telling and strong continuity with its world.... Whatever engages our attention due to its own dignity does so in important part as an embodiment and disclosure of the world it has emerged from. (1995, pp. 38-40)

I have shown how we can discover plants to be such genuine presences that have a "telling and strong continuity with [their] world" and reveal themselves as "an embodiment and disclosure of the world [they have] emerged from." Through such discovery you develop, to use Goethe's expression again, a kind of "organ of perception" for genuine presences because you have engaged in them in one sphere of life in a concrete and experiential way. Then you see that all natural phenomena can become genuine presences that allow meaning-filled experiences to take place; we can speak of encounter-based learning.

Certainly, one can only wish that all people would have the opportunity to quietly experience a sunrise or a sunset, to watch the glistening surface of the ocean and the waves rhythmically pounding the beach, to sit in an old growth forest, to see a flock of migrating birds, and so on. To participate in such presences means to be touched and nourished—for a moment to feel wonder and awe that is an expression of the world in its awesomeness speaking to us. There is no question of the intrinsic value of such encounters. The nourishing contemplation of natural phenomena creates a bond between us and other genuine presences in the world. It also provides a basis for recognizing what is fragmented and isolated, because when you have met things that have a "telling and strong continuity" with their context, you notice when such conditions are not present.

This bond is also created in active play and work (see, for example, Crawford, 2009; Louv, 2006; Sobel, 2008). Let me relate a story that says more than any discursive treatment of this subject could—and also raises wonderful new questions.

I was traveling with my three children on Interstate 80 in Nebraska. We had been through Iowa and seen the endless fields of soybeans and corn. Now corn dominated. At one point the sea of green was interrupted. Huge feedlots with thousands upon thousands of cattle mulling around on barren, muddy ground stretched out before our eyes. The stench was not to be missed. I had told my children earlier that most of the soybeans and corn we'd seen would be fed to cattle. Now they saw them. Where did all the cattle come from? From ranches farther west, where they had roamed upon huge expanses of grassland. They were now at the feedlots for "finishing"—fattening up to be sold to the large meatpacking companies that provide beef for America's consumers.

My oldest daughter, Christina, was fourteen at the time. After seeing the feedlots she made a decision: "Dad, I'm becoming a vegetarian." And she did. Her decision was not based so much on "being grossed out," which all three of my children were. Her concern was for the animals. How can human beings treat other creatures in this way? Cattle have done nothing to deserve being packed into such confinement and fed unnatural high-protein feed just to fatten them up for us. It's just not right, she felt, and from one day to the next she stopped eating meat.

I admired her decision, a decision based on her concern for other creatures. The rest of us remained meat eaters, but from that point on we were much more conscientious about buying only organic beef and, best of all, local grass-fed beef, which we could know had not come from a feedlot animal.

Christina's connection to cattle did not begin at age fourteen. When she was nine and ten years old, during the summer months and on weekends throughout the whole year, she and her younger sister, Franziska, would walk a half mile down to the local organic/biodynamic farm and help with the milking of the sixty-head herd. Christina loved being in the big barn with those big-bodied warm creatures. She fed them hay and curried them. She helped herd them into the barn for milking and cajoled them out again. While a farmer milked the cow, she would hold the tail so it would not swish in his face. The farmers told us later that Christina loved to lie on a cow's back or just stand between two cows that were feeding and feel the warmth emanating from their bodies.

I thought of this connection later on when I pondered why the experience of seeing the feedlots was so strong for her. It was so strong because she knew those beings over there. They were not anonymous four-legged things. Christina knew cows and they had a place in her heart. And when we saw them in a "cow-unworthy" situation in the Nebraskan feedlots, she made a decision that changed her, and her family's, life.

Christina was fortunate to have many experiences of commanding presences in her childhood, and her work with cows at a young age was formative. It did not mean she was preparing to be a farmer; no, she was experiencing and interacting with presences—the morning twilight and cold on her walk to the farm, the cows, the farmers, the cats that roamed the barn. These experiences helped her to become rooted in the world. Later, when she experienced the crassly decontextualized reality of cattle in the feedlots, her being was shaken. After all, cattle had become part of her being. She drew consequences and acted.

I am not saying there is some simple cause-and-effect relation between her childhood experiences at the farm and her response to the feedlots. What did it mean that Christina was nine and ten years old when she worked with cows? Or that at the time of our trip she was opening out into the world in a way that was new? At any rate, the constellation of relationships that had grown in her and that came to meet her on this trip led to a significant life change for her. Interestingly, it was around this time that she decided she wanted to become a social worker. Since she was young she had loved books that portrayed the repressed, the underdogs, those who are discriminated against. Put abstractly, she had a sense for social justice. The things she sought out in the world and the experiences from which she extracted most meaning were ones that helped her more deeply understand this area of life. So what we see is a unique individual on her path of growth and development, and we see that this development is crucially furthered by the interaction with genuine presences.

But what about all the children who do not grow up in environments providing the genuine presences of nature and the possibilities to interact with them, encouraged by the adults in their lives? On the one hand, there are ample genuine presences of nature in urban environments. We often just don't pay attention to them. When Aldo Leopold maintains that "the weeds in a city lot convey the same lesson as the redwoods," he has been able to see those weeds as genuine presences and not merely as poor specimens confined to and damaged by an urban environment (Leopold, 1989, p. 174). There are also clouds to be seen, rain splattering on the sidewalk, the wind rushing through the streets creating swirling eddies of leaves and dust, the cool shade of the trees in the small park, and the birds nesting on the "cliffs" of the apartment building. It's all a matter of whether we are open to and notice such genuine presences. When we begin to do so, more will show themselves. The "we" I mean here are the adults who orchestrate the learning environments for children.

But there's more. Many years ago I came across a remark by the Swiss educator Jakob Streit that children who live in nature-poor urban environments should have all the more opportunity to hear fairy tales (Streit, 1996). At a visceral level this made sense to me, since I had witnessed how engrossed my own and other children were when they listened to fairy tales and other stories. There was no question that these stories are nourishment for the growing soul. I later realized that in many ways stories in human culture resemble organisms in the natural world. Each story has a particular gestalt that is dynamic and membered. There is a flow of development in a story, characters who express different qualities, go through trials and experience resolutions. Real stories are also, like living organisms, always greater, deeper, and richer than any particular interpretation of them. In this way they allow for the great variety of different individual soul configurations and trajectories to enter them and find meaning. Stories are genuine presences.

Or think of art. Works of art are genuine presences. In listening to music, reading literature, or admiring a sculpture, we participate in the presences that other people have produced. But there is also the creative process of production itself. And here I don't want to understand artistic process in any narrow terms, but in the broad sense of bringing something meaning-filled into being. In bringing something into being we are using our own bodies (singing a song, kneading bread, wielding a brush) and working with a variety of media (air, wood, musical instrument, pigments, flour, metal). These provide both the possibilities for creation and resistance to the process. The air, the wood, or the clay are their own presences with their own intrinsic lawfulness. We have to get to know them in the doing. The learner thereby comes into interaction with realities that often demand flexibility and practice to adapt one's capacities to the demands of the medium. The media themselves are genuine presences. Gradually, as we practice and gain some degree of mastery, a medium can become expressive of something that has its own integrity and beauty—the work of art or the useful object. In this way human beings bring new genuine presences into the world.

The El Sistema music program for disadvantaged youths, which began in Venezuela over thirty years ago, shows the power of artistic engagement in transforming human life (http://elsistemausa.org). Here children enter a new life situation that is full of new opportunities for them and provides a rich learning environment. They learn through their own practice and experience accomplished artists playing music. They work together in choirs and orchestras to create music, which among other things is an eminent practice in contextual awareness. The children are guided by artist teachers infused with their concern for the children and their love of art. All this creates a vibrant learning community woven out of the fabric of interacting genuine presences (the engaged teachers, the learners, the instruments, the group productions, the music, etc.).

The spectrum and variety of genuine presences in the world is vast and includes human beings and at least some of our creations. From an educational perspective, the essential task is to orchestrate learning situations so that increasingly genuine presences can become educators. We have done a good job of excluding them from institutionalized learning environments. Wouldn't it be a wonderful task for parents and teachers to work on breathing life into schools by bringing genuine presences into education? Of course this would also mean finding ways to shed (or to circumvent) the many solidified structures and expectations that form the social and political concrete of institutionalized education. It's a tragedy that national governments have combined into one monolithic mission two separate tasks—that of guaranteeing access to affordable education for all, which is the rightful responsibility of the state, and that of discerning the why, how and what of education, which should be shaped by educators (and I include parents as educators) and learners and not by bureaucrats or testing companies. Because of this unhealthy fusion of two distinct tasks, most of the creative thinking and doing in education happens outside the monoculture of public education. (Which isn't to say that individual teachers don't undermine the system and do excellent work.)

The Developing Human Being

What we teach and how we educate should be derived only from our knowledge of the becoming human being and his or her individual potentials. A true science of the human being should be the basis of education and instruction. We shouldn't ask: What does a human being need to know and to master for society as it exists? Rather, we should ask: What are a human being's predispositions and potentials for development? Then it will be possible for each generation to infuse ever new impulses into society. Then what flows out of whole human beings can live in society rather than a new generation becoming a result of what existing society wants to make out of it. (Steiner, 1972, p. 26; transl. by Craig Holdrege)

By asking (in 1919), "What are a human being's predispositions and potentials for development?" Rudolf Steiner, the founder of the Waldorf school movement, places human beings as genuine presences at the center of education. In the recognition that human beings are developing beings lies the source of hope that each generation may bring dispositions and potentials to create a future that is not simply a consequence of the past. The more we know about the developing human being—the more we are engaged in the ongoing task of developing what Steiner calls a "true science of the human being"—the more we can craft learning environments in which we acknowledge the child's past and present ("predispositions") and at the same time work to do justice to what we perceive as the germinal potential in a person.

As John Dewey wrote,

It is no reflection upon the nutritive quality of beefsteak that it is not fed to infants. It is not an invidious reflection upon trigonometry that we do not teach it in the first or fifth grade of school. It is not the subject *per se* that is educative or that is conducive to growth. There is no subject that is in and of itself, or without regard to the stage of growth attained by the learner, such that inherent educational value can be attributed to it. (Dewey, 1997, p. 46)

How often, as a mentor of new teachers, did I find myself asking, when they told me what they planned to teach, "Why do you think the students should learn this?" I wasn't satisfied if the only reason they could give was that someone told them that it is what "one" is supposed to teach ("it's in the curriculum"). I asked them to ponder other questions: How is the subject matter and the way you are working with the material and the students helping them to develop as human beings? Is it fostering careful observation and thinking? Is it leading them into a process whereby they learn the delights and pitfalls of forming judgments? Is there something that can touch them deeply? Does your plan take into account questions that the students have? How does it relate to burning questions of the time? And, if I were in this mentor position today, I would ask: where are the genuine presences in your educational process and how are you helping the students to meet them? The never-ceasing task as an educator becomes: can I continue learning about the developing human being so that I can better discern what kinds of experiences are fruitful and further growth for different individuals, at different ages, and within the cultural and historical context I'm working in?

Out of his deep concern for the future of the earth, David Orr suggests that "no student should graduate from any educational institution without a basic comprehension of things like the following: the laws of thermodynamics, the basic principles of ecology, carrying capacity, energetics, least-cost/end-use analysis, limits of technology, appropriate scale, sustainable agriculture and forestry, steady-state economics, and environmental ethics" (2004, p. 14). As an educator, Orr knows that simply adding this new content to traditional ways of teaching will not do. In a sense we'd be substituting one set of abstractions for another. Is it really so different to learn that "DDT is good for me," as was the slogan in the 1950s when classrooms in the United States were being sprayed with this pesticide, or today that "CO_2 causes global warming," if both are taught as pieces of information to "know"? Of course it is, but I think you can see the point. A course concerned with sustainability or the environment needs to do much more than transmit information. In considering Orr's list of topics we need to ask many questions if we are serious about the educational process and the deeper foundations of sustainability. What experiences can be the basis of learning these concepts? How can they be brought in meaningful ways in relation to the developing human being? In what genuine presences are these high-level abstractions rooted? What kinds of encounters and learning will take students beyond knowing intellectually about the interconnectedness of all things (which is important) to an existentially felt connectedness? These are the kinds of questions that inform living educational thinking.

Speaking in 1919 to the people who would become the teachers at the first Waldorf school, Steiner emphasizes the need to develop ideas and concepts in such a way that they can grow as the children grow. I quote a lengthier passage because, I believe, it captures something that is no less essential nearly a century later:

> What does a concept need to be like when children learn it? It must be living if children are to live with it. Children must live, and, therefore, concepts must also live. If you inoculate nine- or ten-year-old children with concepts that remain the same when those children are thirty or forty years old, then you inoculate them with the corpses of concepts, because the concept does not evolve as the children develop.... When do you inoculate them with dead concepts? When you continually give them definitions, when you say, "A lion is ..." and have them memorize this, then you inoculate children with dead concepts. You assume that when the children are thirty years old, they will have exactly the same concepts you now teach. This means that continual defining is the death of living instruction. Then, what must we do? In teaching, we should not define, we should attempt to characterize. We characterize when we look at things from as many points of view as possible.... We must try to portray animals from different standpoints in the various areas of instruction, for example, from the standpoint of how people came to understand an animal, how people use the animal's work and so forth.... Do not simply

describe a squid and then again later, a mouse, and again later human beings (during different classes), but rather place the squid, mouse and human being next to each other and relate them to one another. In this case, these relationships are so manifold that no single definition emerges, but rather a portrayal. From the very beginning appropriate instruction works not toward definitions, but toward characterization. (Steiner, 1996b, pp. 153–154; translation modified by Craig Holdrege)

The dead concept "is," the living concept can still develop; it is not finished. The living idea of an animal is not the list of its characteristics, but rather the understanding of an animal that emerges when you study it by seeing its characteristics in relation to each other and learn about it in relation to other animals, to its environment, and to human activity. What arises is a many-sided, open-ended understanding of the animal within the larger fabric of reality.

This kind of learning takes time. You cannot rush from one animal to the next. You need to let it become for the children a presence, one illuminated through other presences. Having dwelled with a few animals is much more than having raced through an overview of the whole animal kingdom. It may at first seem paradoxical that precisely when we do not try to cover everything ("completeness") and instead dive into select exemplars, the quality of wholeness can be present. Educator Martin Wagenschein puts it this way:

We recommend the *courage to leave gaps*, which means the courage to be thorough and to dwell intensively on selected topics. So instead of evenly and superficially walking through the catalog of knowledge, step-by-step, we exert the right—or fulfill the duty—to really settle in somewhere, to dig in, to grow roots and take root.... The particular aspect we delve into is not a stage in a process, but a mirror of the whole. Why? The relation the particular has to the whole is not that of a part, step, or preamble; it is a center of gravity. It may be only one, but it carries the whole in it. This single aspect is not an element in a process of accumulation, rather, it carries and illuminates. It is not a stage in a progression, but it works like a spotlight. It affects things that are distant yet related through resonance. (Wagenschein, 2009)

One of the wonderful challenges for a teacher in planning learning experiences is to set aside the nagging catalog of topics that "must" be covered and to consider what topics allow the kind of settling in that Wagenschein means.

As Wagenschein saw, dwelling on a topic "has direct consequences for the schedule. Exemplary teaching doesn't fit in a chopped-up schedule with forty-five minute periods. It needs block scheduling so that we can work on the same theme for two hours every day. That way teaching and learning find a way into the hearts of both students and teachers—and will be working there day and night" (Wagenschein, 2009).

Adult Education at The Nature Institute

At The Nature Institute we want to avoid the disjointedness you find in so many educational programs. But we do not want only to avoid something "bad"; we want to provide a flow and rhythm that has life to it. Since 2002 we have held adult-education courses and workshops. Nearly half of the attendees have been educators (from kindergarten teachers to college professors). Most of our courses are weeklong intensives. Here is one example of a summer intensive with its daily schedule:

Morning sessions
    Flexible thinking—thought exercises in geometric transformation
    Plant study and reflections on method

Afternoon sessions
    Observational work in small groups
    Artistic activity (drawing, painting, or sculpting)

End-of-day forum

The morning and afternoon sessions are each about one-and-a-half to two hours long. Three different teachers are involved and take responsibility, respectively, for crafting geometry, plant study, and artistic work.

We begin each morning with exploring geometric forms and transformations. In this work participants become active in their thinking and more aware of their own thinking activity. They get to know, for example, the qualities of circles, triangles, and other geometric forms as their own kind of genuine presences. And through projective geometry (a non-Euclidean geometry), they begin to form the idea of the infinitely distant—a challenging, stimulating, and mind-stretching experience. This intensive inner work flows into the construction of geometrical drawings.

We then turn our attention to sense experience and the presences that surround us in nature. We carry out plant observations as I described in the previous chapters and begin forming experience-based concepts. In reflecting on these activities we move into considerations of method. Important here is the oscillation between observation and thinking, so that the one activity fructifies the other: if we stay immersed in observation we tend to lose ourselves in details, while if we stay in the reflective mode we tend to lose ourselves in generalities. The way to deal with these dangers is not to find a "middle ground" that has a bit of each (the weak middle), but rather to move energetically from pole to pole. This is, in the realm of inquiry, a form of breathing.

After the lunch break the participants divide into groups of three or four. They are given an observational task: For example, to observe tall buttercup—following and describing the transition from flower bud to open flower to fruit formation; or, to observe tall buttercup specimens that grow in different environments and to compare them with each other. This work in small groups gives participants the opportunity to interact closely with others. It also allows them to practice on their own what was introduced in the morning session.

Drawing allows for quiet focus and individual exploration. Each person works, for example, on attending to forms or a scene as they appear in different gradations of brightness and darkness, and sketches without introducing imagined outlined boundaries. Each day concludes with an open forum. Participants ask questions, they share impressions, and we enter conversation. Sometimes we look back on the day and briefly describe the activities we have carried out.

In these intensives each person performs a small task: putting out materials, sweeping the floor, emptying the compost, washing dishes after a snack, etc. In these small but significant tasks we take responsibility for the space in which we work. It is striking how such activity connects us all with the place.

The days are intense, but because there is a variety of activities and ample time for each activity, people are usually not tired at the end of the day. It helps that evenings are free and people can be on their own and rest or get together for dinner and conversations.

Clearly, the contents of the sessions are different from each other and engage participants differently. This variety could theoretically be experienced as a kind of potpourri, but comments by participants show it is not: "I felt the course was indeed an organic whole. Each part, so different, nevertheless fed into the other, balancing, encircling, and complementing each other. I came away each day knowing my mind had been stretched and opened, my perceptions deepened and sharpened, my soul enriched and fed." Or: "It had a rhythm and balance that was very satisfying. Perhaps you could say it was, in some way, a reflection of nature itself! The combination of studies was very effective." It is a high praise for us to have a course compared to an "organic whole" or "a reflection of nature itself."

Evidently, something living and whole has been created in the way courses unfold. Naturally, not everyone feels this way. But we know that many people do, since such comments repeatedly appear in course evaluations and in a survey of past course participants (see Holdrege, 2010). What stands out for many participants is the experiential and integrative quality of learning. A course—even if it has lasted only a week—can have lasting effects on how participants view things and positively impact daily and professional life. For example, people from a variety of vocations (health professionals, educators, farmers, artists) report how their ability to perceive more openly and carefully and to notice contexts in which events occur was heightened.

Part of the reason why the different topics and activities enhance each other is that each of us who guides a session also participates in the whole course. We don't teach a session and leave. Therefore, each of us relates to what has been done before. We alter our plans to take up and deepen what has already developed. This approach is enlivening for all involved, and learning becomes a responsive, interactive, and dynamic process, leading to discoveries and insights that none of us would have hoped for at the beginning of the week. Numerous times the work in projective geometry has helped me and also the group come to new and unexpected insights in the plant study. Moreover, since each of the teachers is working from a phenomena- and experience-based perspective, a common thread weaves between the contrasting contents: "The form of the course was a wonderful model for wholeness in thinking: the geometry, plant study, and drawing triad, and most importantly how the three teachers worked together—that too came across as a whole and supported the content."

The phenomenological way of working provides a context in which people feel safe and free to contribute: we stand before a world waiting to disclose itself, and each of us has a unique perspective that allows different aspects of the world to come to expression (see also chapter 2, pp. 47–48). We are learning through one another about the world. A high school teacher, who returned to graduate school for a year, writes in a survey response, "Although 'learning from experience' and 'professional learning communities' are trendy catch phrases in academia, I have found them to be poorly understood and rarely practiced. What The Nature Institute provides, an opportunity to rigorously learn from an experience of nature in collaboration with others, is, I believe, one of a kind." Such a learning community forms when adults learn and work together in a respectful, engaged, and mutually fructifying way. What arises is a social genuine presence.

Who Are You?

I want to come back to the future. If you reflect on some of the most important events in your life—ones that evoked growth and development, that allowed something *new* to happen—they were probably not events that school explicitly prepared you for. Were you taught how to find your life's partner in school or prepared for that moment in your life when your first child is born and your life radically changes? Even if someone had told you about the transforming effects of such an event, the actual experience is something wholly other than hearing about it.

Or think of cultural change. Who would have imagined fifty years ago that the book of an unassuming scientist would help ignite a new kind of environmental awareness? I mean Rachel Carson and her book *Silent Spring*. Which educational institutions in the late 1950s and early 1960s were preparing students to be receptive to what Rachel Carson presented? The reception of her book was a surprise, unexpected and exceedingly important.

Such is the unpredictable nature of the future. Another aspect of the future shows itself when we attend to the fact that Carson's book fell on fertile soil in the minds of her readers. Evidently, the book spoke to latent fears and longings, to an underlying openness in people to view human action within the context of planetary ecology. The book helped to awaken these latent potentials, and they became active and took on form in the environmental movement, a new development within human culture. In this way the future, viewed as an emerging force of change and redirection, works into the present. In such reorientation, the feeling of hope is essential. Through hope we are not limited by what is, but dwell in the possibility of what can become. It nurtures the openness that is a prerequisite, as Steiner puts it, "for each generation to infuse ever-new impulses into society."

A few years ago I was asked to give a talk at a high school graduation ceremony in a Waldorf school. In considering what I would say in this brief speech, I knew for sure that I didn't want to say, "I hope the school has prepared you well for college or for life." But I didn't know what to say. I lived—and agonized—in this uncertainty for quite some time. Then, in one moment, it came to me. I needed to say: "My hope is not that the school has prepared you for present-day culture and its existing structures and processes. Rather, my hope is that you have been educated in such a way that the world is not prepared for you. I hope you have not been hindered and that you may even have been nurtured and encouraged to develop ideas and to do things that no one expects—not in order to be different, but because you sense what needs to happen." I added, "Don't listen to people who tell you, when you are following a yearning or birthing an idea, that 'it can't be done.'"

As educators we are already doing something important when we bring learners (ourselves included) into interaction with genuine presences through which each individual in his or her own way can grow. When learning is a discovery process it is open-ended and allows for individual questions and perspectives to come into play. But there is an additional underlying attitude of mind that, in my experience, plays an important role in the learning process.

I can best express this attitude of mind through questions. When I was teaching high school, in my work with the students I tried to keep these questions alive: Who are you? I am working with you on a daily basis and yet I don't know you. What is it that you want to realize in your life?

Neither I nor the student can answer these questions. If we could, it would mean there would be no development. Everything would be clear.

Through an ever-renewed effort to engage this questioning, searching attitude of mind and to work with the students out of it, something new and essential arises in the learning community. What happens is that the students become "large"; that is, I don't see them just as adolescents now with their quirks, gifts, and difficulties, but as participants within a developmental stream of human life. Moreover, I acknowledge in the students a dimension of inner depth—a realm out of which their individual questions and strivings arise. This realm remains hidden for me if I get caught up in the outer trappings of adolescence. I know that in each student something wants to grow like the growing point of a plant—vulnerable, tender, and full of life. I don't want to crush that! I'm dealing with a kind of "holy of holies" in each student that warrants deep respect. It needs protection, and it needs soul space and biographical time to develop.

In this attitude of mind I become a listener. Can I hear what it is that you are really asking—and listen through the pointed question or the cold logic with which you argue? I'm trying to hear the meaning or intent that arises out of the deeper, hidden source that speaks "between the lines" in word, gesture, and action. And inasmuch as I do hear something, my inner response is: how can I serve what you are saying through my work with you? This is, to state the relation differently, the attitude of teacher as a midwife, who helps give birth to that which wants to come into the world and thrive.

In my experience, students notice whether you are working out of such an attitude—which is not explicit but implicit in all the smaller and bigger interactions that occur. It provides a kind of fertile ground out of which manifold learning experiences arise.

Sustainability and Education

In a similarly open and concerned attitude of mind we can look to the whole planet. The planet is nothing static and fixed. Life and conditions on earth have evolved and will evolve in ways we cannot predict. But there is no doubt that we are participants in this evolutionary process. It is inimical to life to look at sustainability as a goal to be achieved. Stephen Rockefeller writes that "sustainability includes all the interrelated activities that promote the long-term flourishing of Earth's human and ecological communities" (cited in Carroll, 2004, p. 2). I agree as long as we realize that this "long-term flourishing" includes evolutionary processes and that it does not mean trying to hold onto everything we value today.

This view may seem to be "dangerous." But if we ignore the transformational, evolutionary nature of life, then our programs will always come up short. It is essential that we model our ways of thinking after living processes, and this has been the main concern of this book. Without living thinking, we won't find ways of acting that support the "long-term flourishing" of the planet. But once we begin to understand life, we also realize that is open-ended. Therefore, in any given conceptual framework or plan of action, we need to be keenly aware of our ignorance and the boundaries of our current insights. Socrates' statement that wisdom consists in awareness of our ignorance has lost no pertinence over nearly two and a half millennia. Awareness of what we do not know makes us more cautious, circumspect, and critical, and infuses us with a readiness to change our ideas and course of action. It is a key component of a real ability to evolve and adapt to changing circumstances.

So at a fundamental level, sustainability hinges on evolving human capacities and the willingness to engage in a never-ending process of human involvement that itself will evolve in as-yet unknown ways in the future. As an attitude of mind, sustainability would entail wholeheartedly embracing an uncertain journey.

This planetary attitude of mind and the developmental attitude of an educator are in fact two aspects of one living way of being. Wherever in the world learners are engaging in some form of exploratory, encounter-based learning with the earth's genuine presences, they are, I believe, developing capacities they need to be creative participants in the planet's healthy evolution. The more this occurs, the more reason we have to hope that society-as-status-quo will not be prepared for what future generations will bring.

Where Do All These Plants Come From?

In the early 1990s I moved to rural upstate New York and began teaching high school at the Hawthorne Valley School, an independent Waldorf school. I was primarily a "block" teacher; blocks were three to four weeks long, and I worked with a class two hours each day during the block. This schedule framework allows you to delve into a topic. During the first years I developed some new block courses. I was grateful to my colleagues for being open to my suggestions so that I was able to explore territory new both to me and to my students. The school is surrounded by a biodynamic farm and woodlands. It's a rich natural and agricultural environment. I developed an ecology course in the eleventh grade that focused on forest, pasture, and meadow communities, and a botany course in twelfth grade that focused on spring wildflowers in different habitats. In both courses we spent much time outside studying plants and habitats.

The botany course was three weeks long near the end of the school year. At the end of the school year shortly before graduation, twelfth graders are not exactly "present" and eager to be in school. This course came right before the students were to present their individual year-long projects and before their stage play that was a kind of culmination of their school experience. So you can imagine that none of the teachers were especially enthralled about the prospect of working with the twelfth graders at this time. Was I naive to think the students would become interested in plants? From the outset I knew that some of my more lofty ideas about botany and plant metamorphosis, things I've discussed in previous chapters, would probably not be part of the course. It wasn't a time for philosophical reflection—something I did a lot of otherwise with eleventh and twelfth graders. It was a time to dive into sense experience, but in a structured way. So I developed the block as a field course, and the plants themselves taught most of the content. We'd go outside nearly every day and observe, describe, and identify wildflowers growing in the different environments around the school.

I was glad—and quite surprised—to see how interested the students became in the plants. Through careful observation, drawing, and comparing, and subsequent conversations about plant structures and functions in relation to their observations, the students began to understand that plants are quite remarkable. By observing many different plants they also began to get a sense for different growth forms, flowering patterns, and the relations of specific species to specific environments.

In one class toward the end of the block we were sitting at the top of a wooded hill studying the wild columbine, a plant that grows on rock outcrops. It was hard not to be drawn to its remarkable dangling and intricate scarlet-red and bright-yellow flowers. While the students were observing, writing, or drawing, one of them asked, "Mr. Holdrege, where do all these plants come from?" From the whole situation, it was clear to me that this was not a question to be answered. Every answer would have fallen flat in light of what, for a moment, this student had inwardly touched. I think I just looked at her and nodded in the inner acknowledgment that I shared the same unanswered question. That was a golden educational moment I cherish to this day.

Something of the normally unmanifest and deep nature of plants had become present in this student's soul and her response to this meeting was wonder and a question. The experience of such a presence cannot be clearly outlined or defined because it is an opening into a reality that has depth and potential—that can still become. The encounter is alive and vital; we touch a common source of becoming in ourselves and in the world.

Plants are remarkable teachers.

Appendix

Root Boxes

Biologist Jochen Bockemühl designed root boxes to allow the observation of the development of the roots during the life cycle of the plant (Bockemühl, 1969). The root boxes are 80 cm high, 23 cm wide and 3 cm deep; they are filled with soil. One side of the box is glass. The box is tilted at an angle so that the roots grow, as they grow downward, toward the glass and then down alongside the glass. Thus they can be observed.

   Roots show positive geotropism, which means they tend to grow toward the earth's center, so when the box is at an angle, they grow perpendicular to the earth's surface and toward the glass. They continue to grow along the glass, making them visible against the inside of the glass. The glass is covered while the plant grows, so that the roots are growing in darkness, and only uncovered to allow them to be traced.

Notes

Chapter 1

[1] What I am calling object thinking has been described, to mention a few examples, as the intellect (Bergson, 1998) or intellectual mind (Bortoft, 1996), onlooker or spectator consciousness (Lehrs, 1958), nonparticipatory consciousness (Berman, 1984), the mechanical view of nature (Merchant, 1983) or the mechanized world picture (Mumford, 1970), fragmentation (Bohm, 2002) or fragmented thinking (Sloan, 1983), reductionism (for a review, see Brigandt & Love, 2008) or the reduction complex (Talbott, 2004a), and positivism (Bhaskar, 1987). Each of these expressions points to different nuances of the worldview, just as my choice of the term object thinking aims to accentuate an essential feature of this frame of mind.

[2] Bortoft (1996) has characterized, in a similar way, the difference between what he calls counterfeit and authentic wholes, while Talbott (2001, 2002) has exposed the reductionism inherent in complexity theory.

Chapter 2

[1] In their book *Presence* Peter Senge and his colleagues describe a first stage or level of knowing as "sensing," which corresponds to the perceptual immersion I have been discussing (Senge, Scharmer, Jaworski, and Flowers, 2004). Then they speak of "presencing," in which we "retreat and reflect" and "perception starts to arise within the living process of the whole" (p. 89). This notion is very similar to the practice of exact sensorial imagination and has been developed further by Otto Scharmer, especially in relation to enlivening perception and insight within organizational and social contexts, in his book *Theory U* (Scharmer, 2007).

Chapter 3

[1] My efforts here and in the following chapter build on Goethe's approach to botany (Goethe, 1989) and its further development by other scientists. For applications of the Goethean approach to botany in the English language, see Bockemühl, 1981; Bockemühl & Suchantke, 1995; Colquhoun & Ewald, 1996; Goethe, 1989; Grohmann, 1989a, 1989b; Hoffmann, 2007; and Suchantke, 2009.

Chapter 5

[1] I have carried on studies on the cow (Holdrege, 2008a), horse and lion (Holdrege, 1998), elephant (Holdrege, 2003), giraffe (Holdrege, 2005a), and sloth (Holdrege, 2008b) and the skunk cabbage (Holdrege, 2000b).

References

Agrawal, A. A. (2005). Natural selection on common milkweed (*Asclepias syriaca*) by a community of specialized insect herbivores. *Evolutionary Ecology Research, 7,* 651–667.

Alberts, B. (2010). An education that inspires. *Science, 330,* 427.

Ashton, P., & Berlyn, G. (1994). A comparison of leaf physiology and anatomy of *Quercus* (section Erythrobalanus-Fagaceae) species in different light environments. *American Journal of Botany, 81,* 589–597.

Barber, J. R., & Conner, W. E. (2007). Acoustic mimicry in a predator-prey interaction. *PNAS, 104,* 9331–9334.

Barfield, O. (1967). *History in English words.* Great Barrington, MA: Lindisfarne Books.

Barnes, J. (1999). *Goethe and the power of rhythm.* Ghent, NY: Adonis Press.

Baum, H. (1948). *Der Fruchtansatz von Asclepias syriaca* [The fruit primordium of *Asclepias syriaca*]. *Plant Systematics and Evolution, 94,* 402–403.

Berenbaum, M. (1993). Sequestered plant toxins and insect palatibility. *The Food Insects Newsletter, 6(3),* 1–10.

Bergson, H. (1998). *Creative evolution.* Mineola, New York: Dover Publications. (This is a reprint of the 1911 translation; the French original was published in 1907.)

Berman, M. (1984). *The reenchantment of the world.* New York: Bantam Books.

Bhaskar, R. (1987). *Scientific realism and human emancipation.* New York: Verso.

Bockemühl, J. (1969). Gartenkresse, Kamille, Baldrian [Garden cress, chamomile, and valerian]. *Elemente der Naturwissenschaft, 11,* 13–28. (This article describes root boxes.)

Bockemühl, J. (1973). Entwicklungsweisen des Klatschmohns im Jahreslauf als Hilfen zum Verständnis verwandter Arten [The development of the field poppy during the year as an aid to understanding related species]. *Elemente der Naturwissenschaft, 19,* 37–56.

Bockemühl, J. (1981). *In partnership with nature.* Junction City, OR: Biodynamic Association.

Bockemühl, J. (1982). Staubblatt und Fruchtblatt [Stamen and pistil]. In W. Schad (Ed.), *Goetheanistische Naturwissenschaft: Botanik* (pp. 115–129). Stuttgart: Verlag Freies Geistesleben.

Bockemühl, J. (1985). Pflanzengestalt und Lichtverhältnisse [Plant form and light relations]. In B. Endlich (Ed.), *Der Organismus der Erde* (pp. 107–117). Stuttgart: Verlag Freies Geistesleben.

Bockemühl, J., & Suchantke, A. (1995). *The metamorphosis of plants.* Cape Town: Novalis Press.

Bohm, D. (1989). *Quantum theory*. New York: Dover Publications, Inc.

Bohm, D. (1996). *On dialogue*. New York: Routledge Classics.

Bohm, D. (2002). *Wholeness and the implicate order*. New York: Routledge Classics.

Bohm, D. (2005a). *On creativity*. New York: Routledge Classics.

Bookman, S. S. (1981). The floral morphology of *Asclepias speciosa* (*Asclepiadaceae*) in relation to pollination and a clarification of terminology for the genus. *American Journal of Botany, 68,* 675–679.

Borgmann, A. (1984). *Technology and the character of contemporary life*. Chicago: The University of Chicago Press.

Borgmann, A. (1995). "The Nature of Reality and the Reality of Nature." In M. E. Soulé and G. Lease (Eds.), *Reinventing Nature* (pp. 31–45). Washington, DC: Island Press.

Bortoft, H. (1996). *The wholeness of nature*. Great Barrington, MA: Lindisfarne Press.

Bortoft, H. (2007). The dynamics of being—Goethe and modern philosophy. Lecture at Ruskin Mill College, September 14; unpublished.

Brady, R. (1998). *The idea in nature*. In D. Seamon & A. Zajonc (Eds.), Goethe's way of science (pp. 83–111). Albany, NY: State University of New York Press.

Brigandt, I., & Love, A. (2008). Reductionism in biology. *Stanford Encyclopedia of Philosophy*. Retrieved on January 3, 2009, from http://plato.stanford.edu/entries/reduction-biology

Brooks, J. R. Meinzer, F. C., Coloumbe, R., & Gregg, J. (2002). Hydraulic redistribution of soil water during summer drought in two contrasting Pacific northwest coniferous forests. *Tree Physiology, 22,* 1107–1117.

Calaprice, A. (Ed.). (2005). *The new quotable Einstein*. Princeton: Princeton University Press.

Caldwell, M. M., Dawson, T. E., & Richards, J. H. (1998). Hydraulic lift: consequences of water efflux from the roots of plants. *Oecologia, 113,* 151–161.

Callicott, J. B. (2002). From the balance of nature to the flux of nature: the land ethic in a time of change. In R. L. Knight and S. Riedel (Eds.), *Aldo Leopold and the ecological conscience* (pp. 90–105). New York: Oxford University Press.

Carroll, J. (2004). *Sustainability and spirituality*. Albany, NY: State Univeristy of New York Press.

Churchland, P. (1988). *Matter and consciousness: a contemporary introduction to the philosophy of mind*. Cambridge, Massachusetts: The MIT Press.

Cohen, J. A., & Brower, L. P. (1983). Cardenolide sequestration by the dogbane tiger moth (Cycnia tenera; Arctiidae). *Journal of Chemical Ecology, 9,* 521–532.

Colquhoun, M., & Ewald, A. (1996). *New eyes for plants*. Stroud, UK: Hawthorn Press.

Crawford, M. B. (2009). *Shop class as soulcraft: an inquiry into the value of work*. New York: Penguin Books.

Cronon, W. (1992). A place for stories: nature, history, and narrative. *The Journal of American History, 78,* 1347–1376.

Dailey, P. J., Graves, R. C., & Herring, J. L. (1978). Survey of hemiptera collected on common milkweed, *Asclepias syriaca*, at one site in Ohio. *Entomological News, 89,* 157–162.

Dailey, P. J., Graves, R. C., & Kingsolver, J. M. (1978). Survey of coleoptera collected on the common milkweed, *Asclepias syriaca,* at one site in Ohio. *The Coleopterists Bulletin, 32,* 223–229.

Damasio, A. (2001). Fundamental feelings. *Nature 413,* 781.

Darwin, C. (1872). *Origin of species,* sixth edition. London: John Murray. Downloaded on January 6, 2010, from http://darwin-online.org.uk/contents.html

Darwin, C. (1979). *Origin of species,* reprint of first edition. New York: Penguin Books. (The first edition of this book was published in 1859.)

Darwin, C. (2005). *The autobiography of Charles Darwin, 1809–1882:* with original omissions restored (N. Barlow, Ed.). New York: W. W. Norton & Company.

Dengler C., & Tsukaya, H. (2001). Leaf morphogenesis in dicotyledons: current issues. *International Journal of Plant Science, 160,* 459–464.

Desmond, A., & Moore, J. (1994). *Darwin: the life of a tormented evolutionist.* New York: W. W. Norton & Company.

Descartes, R. (1968). Meditations. In W. Kaufmann (Ed.), *Philosophic classics: Bacon to Kant* (pp. 22–63). Englewood Cliffs, NJ: Prentice-Hall, Inc. (Descartes' Meditations were originally published in 1641.)

Dewey, J. (1997). *Experience and education.* New York: Touchstone. (This book was originally published in 1938.)

Duffey, S. S., & Scudder, G. G. E. (1972). Cardiac glycosides in North American Asclepiadaceae, a basis for unpalatability in brightly coloured hemiptera and coleoptera. *Journal of Insect Physiology, 18,* 63–78.

Duffey, S. S., Blum, M. S., Isman, M. B., & Scudder, G. G. E. (1978). Cardiac glycosides: a physical system for their sequestration by the milkweed bug. *Journal of Insect Physiology, 24,* 639–645.

Dussourd, D. E. (1999). Behavioral sabotage or plant defense: do vein cuts and trenches reduce insect exposure to exudate? *Journal of Insect Behavior, 12,* 501–515.

Ellul, J. (1990). *The technological bluff.* Trans. Geoffrey W. Bromiley. Grand Rapids, MI: Eerdmans.

Emerson, R. W. (1990). *Essays: first and second series.* New York: Vintage Books.

Erdman, J., Beirer, T., & Gugger, E. (1993). Absorption and transport of carotenoids. In L. Canfield, N. Krinsky, & J. Olson (Eds.), *Carotenoids in human health* (pp. 76–85). New York: New York Academy of Sciences.

Eschrich, W., Burchardt, R., & Essiamah, S. (1989). The induction of sun and shade leaves of the European beech (*Fagus sylvatica L.*): anatomical studies. *Trees, 3,* 1–10.

Farrell, B. D., & Mitter, C. (1998). The timing of insect/plant diversification: might *Tetraopes* (Coleoptera: Cerambycidae) and *Ascelpias* (Asclepiadaceae) have co-evolved? *Biological Journal of the Linnean Society, 63,* 553–577.

Fordyce, J. A., & Malcolm, S. B. (2000). Specialist weevil, Rhyssomatus liineaticollis, does not spatially avoid cardenolide defenses of common milkweed by ovipositing into pith tissue. *Journal of Chemical Ecology, 26,* 2857–2874.

Fox, C. W., & Caldwell, R. L. (1994). Host-associated fitness trade-offs do not limit the evolution of diet breadth in the small milkweed bug Lygaeus kalmii (Hemiptera: Lygaeidae). *Oecologia, 97,* 382–389.

Fritz, R. S., & Morse, D. H. (1981). Nectar parasitism of Asclepias syriaca by ants: effect on nectar levels, pollinea insertion, pollinaria removal and pod production. *Oecologia, 50,* 316–319.

Gadamer, H.-G. (1977). *Philosophical hermeneutics.* Berkeley: University of California Press.

Gädeke, M. (2000). Goethes Urpflanze und der "Bauplan" der Morphologie [Goethe's archetypal plant and the bauplan of morphology]. In P. Heusser (Ed.), *Goethes Beitrag zur Erneuerung der Naturwissenschaften* (pp. 107–129). Bern: Verlag Paul Haupt.

Gardner, G., & Halweil, B. (2000). *Underfed and overfed: the global epidemic of malnutrition* (Worldwatch paper 150). Washington, DC: Worldwatch Institute.

Goethe, J. W. (1973). *Goethes Werke, Hamburger Ausgabe, Band XII* [Goethe's works, Hamburg edition, volume 12]. Munich: Verlag C. H. Beck. (All English language quotations from this volume that appear in the text have been translated by Craig Holdrege.)

Goethe, J. W. (1982). *Italian journey.* San Francisco: North Point Press.

Goethe, J. W. (1989). *Goethe's botanical writings* (B. Mueller, Transl.). Woodbridge, CT: Ox Bow Press.

Goethe, J. W. (1995). *The scientific studies* (D. Miller, Ed. & Transl.). Princeton: Princeton University Press.

Goethe, J. W. (2002). *Goethes Werke, Hamburger Ausgabe, Band XIII* [Goethe's works, Hamburg edition, volume 13]. Munich: Verlag C. H. Beck. (All English language quotations from this volume that appear in the text have been translated by Craig Holdrege.)

Goethe , J. W. (2010). The experiment as mediator of object and subject (C. Holdrege, Transl). *In Context, 24,* 19–23. (Goethe wrote this essay in 1792.)

Golley, F. B. (1998). *A primer for environmental literacy.* New Haven: Yale University Press.

Goto, K, Kyozuka, K., & Bowman, J. (2001). Turning floral organs into leaves and leaves into floral organs. *Current Opinion in Genetics and Development, 11,* 449–456.

Gould, S. J., & Lewontin, R. C. (1979). The spandrels of San Marco and the Panglossian paradigm: a critique of the adaptationist programme. *Proc. R. Soc. London B, 205,* 581–598.

Grohmann, G. (1989a). *The Plant,* vol. 1. Junction City, OR: Biodynamic Farming and Gardening Association.

Grohmann, G. (1989b). *The Plant,* vol. 2. Junction City, OR: Biodynamic Farming and Gardening Association.

Hanson, H. (1917). Leaf-structure as related to light. *American Journal of Botany, 4,* 533–560.

Harlen, V. (Ed.). (2005). *Wert und Grenzen des Typus in der botanischen Morphologie* [Value and boundaries of the archetype in botanical morphology]. Nümbrecht: Martina-Galunder-Verlag.

Harding, S. (2006). *Animate earth.* White River Junction, VT: Chelsea Green Publishing.

Hartman, F. A. (1977). The ecology and coevolution of common milkweed (*Asclepias syriaca,* Asclepiadacieae) and milkweed beetles (*Tetraopes tetrophthalmus,* Cerambycidae). Ph.D. Thesis, University of Michigan, Ann Arbor.

Heisenberg, W. (1947). *Wandlungen in den Grundlagen der Naturwissenschaften* [Transformations in the Foundations of Natural Science, 7th edition]. Stuttgart: Hirzel.

Heisenberg, W. (1962). *Physics and philosophy.* New York: Harper Torchbooks.

Heitler, W. (1973). Vom Wesen der Quantenchemie [The nature of quantum chemistry], *Phys. Bl., 29,* 252 and 256.

Helmus, M. R., & Dussourd, D. E. (2005). Glues or poisons: which triggers vein cutting by monarch caterpillars? *Chemoecology, 15,* 45–49.

Hoffmann, N. (2007). *Goethe's science of living form.* Ghent, NY: Adonis Press.

Holch , A. E. (1931). Development of roots and shoots of certain deciduous tree seedlings in different forest sites. *Ecology, 12,* 259–298.

Holdrege, C. (1996). *Genetics and the manipulation of life.* Great Barrington, MA: Lindisfarne Press.

Holdrege, C. (1998). Seeing the animal whole: the example of horse and lion. In D. Seamon & A. Zajonc (Eds.), *Goethe's way of science* (pp. 213–232). Albany, NY: State University Press of New York Press.

Holdrege, C. (2000a). Where do organisms end? *In Context, 3,* 14–16.

Holdrege, C. (2000b). Skunk cabbage (*Symplocarpus foetidus*). *In Context, 4,* 12–18. Available online: http://natureinstitute.org/pub/ic/ic4/skunkcabbage.htm.

Holdrege, C. (2003). *The flexible giant: seeing the elephant whole.* Ghent, NY: The Nature Institute.

Holdrege, C. (2005a). *The giraffe's long neck: from evolutionary fable to whole organism.* Ghent, NY: The Nature Institute.

Holdrege, C. (2005b). The forming tree. *In Context, 14,* 18–22.

Holdrege, C. (2008a). The cow: organism or bioreactor? In C. Holdrege & S. Talbott, *Beyond biotechnology: the barren promise of genetic engineering* (pp. 111–122). Lexington: The University Press of Kentucky.

Holdrege, C. (2008b). What does it mean to be a sloth? In C. Holdrege & S. Talbott, *Beyond biotechnology: the barren promise of genetic engineering* (pp. 132–153).

Lexington: The University Press of Kentucky. Available online: http://www. natureinstitute.org/nature/sloth.htm

Holdrege, C. (2009). Evolution evolving. *In Context, 21,* 16–23.

Holdrege, C. & Talbott, S. (2008). *Beyond biotechnology: the barren promise of genetic engineering.* Lexington: The University Press of Kentucky.

Holdrege, C. (2010). The effects of Nature Institute courses: results of a survey. Available online: http://www.natureinstitute.org/educ/NISurvey2009.pdf

Homer (1967). *The odyssey of Homer* (R. Lattimore, Transl.). New York: Harper.

Husserl, E. (1993). *Logische Untersuchungen* (Bd. II, Teil 1) [Logical Investigations (Vol. II, Part 1)]. Tübingen: Max Niemeyer Verlag.

Hutchinson, C. A., Peterson, S. N., Gill, S. R., Cline, R. T., White, O., Fraser, C. M., et al. (1999). Global transposon mutagenesis and a minimal mycoplasma genome. *Science, 286,* 2165–2169.

Jablonka, E., & Lamb, M. J. (2006). *Evolution in four dimensions.* Cambridge, MA: The MIT Press.

James, W. (1950). *The principles of psychology* (volume 1). Mineola, NY: Dover. (This book was originally published in 1890.)

Jennersten, O., & Morse, D. H. (1991). The quality of pollination by diurnal and nocturnal insects visiting common milkweed, *Asclepias syriaca. American Midland Naturalist, 125,* 18–28.

Jones, C. (1995). Does shade prolong juvenile development? A morphological analysis of leaf shape changes in Cucurbita argyrosperma subsp. Sororia (Cucurbitaceae). *American Journal of Botany, 82,* 346–359.

Jones, C. (1999). An essay on juvenility, phase change, and heteroblasty in seed plants. *International Journal of Plant Science 160* (6 Suppl.), S105–S111.

Kant, I. (1951). *Critique of judgment.* New York: Hafner Press.

Kant, I. (1965). *Critique of pure reason.* New York: St. Martin's Press.

Kaplan, D. (2001). The science of plant morphology: definition, history, and role in modern biology. *American Journal of Botany, 88,* 1711–1741.

Kauffman, S. (2008). *Reinventing the sacred: a new view of science, reason, and religion.* New York: Basic Books.

Kephart, S. R. (1987). Phenological variation in flowering and fruiting of Asclepias. *American Midland Naturalist, 118,* 64–76.

Kirchoff, B. (2001). Character description in phylogentic analysis: insights from Agnes Arber's concept of the plant. *Annals of Botany, 88,* 1203–1214.

Kranich, E.-M. (2007). Goetheanism—its methods and sinificance in the science of the living. *Archetype, 13,* 12–24.

Krugman, P. (2008). Life without bubbles. *New York Times,* December 22. http://www.nytimes.com/2008/12/22/opinion/22krugman.html

Kuhn, T. S. (1996). *The structure of scientific revolutions* (3rd ed.). Chicago: University of Chicago Press.

Kuhn, T. S. (2002). *The road since structure.* Chicago: Chicago University Press.

Laland, K. N., & Sterelny, K. (2006). Perspective: seven reasons (not) to neglect niche construction. *Evolution, 60,* 1751–1762.

Larcher, W. (2003). *Physiological plant ecology.* Berlin: Springer Verlag.

Lehrs, E. (1958). *Man or matter.* London: Faber and Faber.

Leopold, A. (1989). *A sand county almanac.* New York: Oxford University Press. (This book was originally published in 1949.)

Leopold, A. (1999). *For the health of the land.* Washington, DC: Island Press. (The quotes are from the essay "The farmer as conservationist," which was originally published in 1939.)

Levins, R., & Lewontin, R. (1985). *The dialectical biologist.* Cambridge, MA: Harvard University Press.

Lichtenthaler, H., Buschmann, C., Döll, M., Fietz, H.-J., Bach, T., Kozel, U., et al. (1981). Photosynthetic activity, chloroplast ultrastructure, and leaf characteristics of high-light and low-light plants and of sun and shade leaves. *Photosynthesis Research, 2,* 115–141.

Lockley, M. G. (2007). The morphodynamics of dinosaurs, other archosaurs, and their trackways: holistic insights into relationships between feet, limbs, and the whole body. *SEPM Special Publication* no. 88 (Society for Sedimentary Geology), pp. 27– 51.

Lovelock, J. (2000). *Gaia: a new look at life on earth.* Oxford: Oxford University Press.

Louv, R. (2006). *Last child in the woods.* Chapel Hill, NC: Algonquin Books.

Lovins, A. (2001). Loaves and fishes (Interview with Susan Witt, September 8, 2001, at the Omega Institute for Holistic Studies). http://www.smallisbeautiful.org/publications/lovins_omega.html

Malcolm, S. B. (1995). Milkweeds, monarch butterflies and the ecological significance of cardenolides. *Chemoecology, 5/6,* 101–117.

Malthus. T. (1999). *Essay on the principle of population.* Oxford: Oxford University Press.

Maslow, A. (1969). *The psychology of science.* Chicago: Henry Regnery Company.

McCollum, E. V. & Davis, M. (1913). The necessity of certain lipins in diet during growth. *The Journal of Biological Chemistry, 15,* 167–175.

Meine, C. (1988). *Aldo Leopold: his life and work.* Madison, WI: The University of Wisconsin Press.

Merchant, C. (1983). *The death of nature.* San Francisco: Harper.

Merleau-Ponty. M. (1969). *The essential writings of Merleau-Ponty* (A. Fisher, Ed.). New York: Harcourt, Brace & World.

Mooney, K. A., Jones, P., & Agrawal, A. A. (2008). Coexisting congeners: demography, competition, and interactions with cardenolides for two milkweed-feeding aphids. *Oikos, 117,* 450–458.

Morse, D. H. (1982). The turnover of milkweed pollinia on bumble bees, and implications for outcrossing. *Oecologia, 53,* 187–196.

Mumford, L. (1970). *The pentagon of power.* New York: Harcourt Brace Jovanovich, Publishers.

Niinemets, U., Kull, O., & Tenhunen, J. (1999). Variability in leaf morphology and chemical composition as a function of canopy light environment in coexisting deciduous trees. *International Journal of Plant Sciences, 160,* 837–848.

Niesenbaum, R. A., Cahill, J. F., and Ingersoll, C. M. (2006). Light, wind, and touch influence leaf chemistry and rates of herbivory in *Apocynum cannabinum* (Apocynaceae). *International Journal of Plant Science, 167,* 969–978.

O'Conner, J., & McDermott, I. (1997). *The art of systems thinking.* London: Thorsons.

Odling-Smee, F. J., Laland, K. N., & Feldman, M. W. (2003). *Niche construction: the neglected process in evolution.* Princeton: Princeton University Press.

Orr, D. (2004). *Earth in mind.* Washington, DC: Island Press.

Paine, J., Shipton, A, Chaggar, S., Howells, R., Kennedy, M. Vernon, G., et al. (2005). Improving the nutritional value of golden rice through increased provitamin A content. *Nature Biotechnology, 23,* 482–487.

Pigliucci, M. (2001). *Phenotypic plasticity.* Baltimore: The Johns Hopkins University Press.

Pleasants, J. M. (1991). Evidence for short-distance dispersal of pollinia in *Asclepias syriaca* L. *Functional Ecology, 5,* 75–82.

Plessner, H. (1975). *Die Stufen des Organischen und der Mensch* [The stages of the organic and the human being]. Berlin: Walter de Gruyter.

Portmann, A. (1967). *Animal forms and patterns.* New York: Schocken Books.

Price, P. W., & Wilson, M. F. (1979). Abundance of herbivores on six milkweed species in Illinois. *American Midland Naturalist, 101,* 76–86.

Prysby, M. D. (2004). In Oberhauser, K. S., & Solensky, M. J. (Eds.), *The monarch butterfly: biology and conservation* (pp. 79–83). Ithaca: Cornell University Press.

Rausher, M. D. (2001). Co-evolution and plant resistance to natural enemies. *Nature, 411,* 857–864.

Richards, R. (2002). *The romantic conception of life: science and philosophy in the age of Goethe.* Chicago: University of Chicago Press.

Riegner, M. (1993). Toward a holistic understanding of place: reading a landscape through its flora and fauna. In D. Seamon (Ed.), *Dwelling, seeing and designing: toward a phenomenological ecology* (pp. 181–215). Albany, NY: State University of New York Press.

Riegner, M. (2008). Parallel evolution of plumage pattern and coloration in birds: implications for defining avian morphospace. *The Condor, 110,* 599–614.

Ruskin, J. (1971). *The elements of drawing.* New York: Dover Publications, Inc. (This book was originally published in 1857.)

Ryle, G. (1949). *The concept of mind.* New York: Barnes & Noble, Inc.

Sage, T. L., & Williams, E. G. (1995). Structure, ultrastructure, and histochemistry of the pollen tube pathway in the milkweed *Asclepias syriaca* L. *Sex Plant Reprod, 8,* 257–265.

Sattler, R., & Rutishauser, R. (1997). The fundamental relevance of morphology and morphogenesis to plant research. *Annals of Botany, 80,* 571–582.

Schad, W. (1977). *Man and mammals: toward a biology of form.* Garden City New York: Waldorf Press.

Scharmer, C. O. (2007). *Theory u.* Cambridge, MA: Society for Organizational Learning.

Schilperoord, P. (2011). *Metamorphosen im Pflanzenreich* [Metamorphosis in the plant kingdom]. Stuttgart: Verlag Freies Geistesleben.

Schlichting, C. D., & Pigliucci, M. (1998). *Phenotypic evolution: a reaction norm perspective.* Sunderland, MA: Sinauer Associates, Inc.

Senge, P., Scharmer, C.O., Jaworski, J., & Flowers, B.S. (2004). *Presence.* New York: Doubleday (Random House).

Sloan, D. (1983). *Insight-imagination: the emancipation of thought and the modern world.* Westport, CT: Greenwood Press.

Sobel, D. (2008). *Childhood and nature.* Portland, ME : Stenhouse Publishers.

Solensky, M. J. (2004). Overview of monarch migration. In K. S. Oberhauser & M. J. Solensky (Eds.), *The monarch butterfly: biology and conservation* (pp. 79–83). Ithaca: Cornell University Press.

Southwick, E. E. (1983). Nectar biology and nectar feeders of common milkweed, Asclepias syriaca L. *Bulletin of the Torrey Botanical Club, 110,* 324–334.

Steiner, R. (1972). *Zur Dreigliederung des sozialen Organismus* [Concerning the threefold social order]. Stuttgart, Germany: Verlag Freies Geistesleben.

Steiner, R. (1983). *The boundaries of natural science.* Hudson, NY: Anthroposophic Press.

Steiner, R. (1987). *Anthroposophie, ihre Erkenntniswurzeln und Lebensfrüchte* (Bibliographie-Nr. 53) [Anthroposophy, the roots of its knowledge and its fruits for life (collected works vol. 53)]. Dornach, Switzerland: Rudolf Steiner Verlag.

Steiner, R. (1996a). *The science of knowing.* Spring Valley, NY: Mercury Press.

Steiner, R. (1996b). *The foundations of human experience.* Great Barrington, MA: Anthroposophic Press.

Steiner, R. (2006). *The philosophy of freedom.* London: Rudolf Steiner Press. (The original German book was first published in 1894; it is available in a number of English translations.)

Sterling, S. (2004). *Sustainable education.* Totnes, UK: Green Books.

Streit, J. (1996). *Warum Kinder Märchen brauchen* [Why children need fairy tales]. Dornach, Switzerland: Verlag am Goetheanum.

Suchantke, A. (2009). *Metamorphosis: evolution in action.* Ghent, NY: Adonis Press.

Talbott, S. (2001). The lure of complexity. *In Context, 6,* 15–19.

Talbott, S. (2002). The lure of complexity (part II). *In Context, 7,* 19–23.

Talbott, S. (2004a). The reduction complex. *NetFuture, 158.* http://www.netfuture.org/2004/Nov0904_158.html

Talbott, S. (2004b). Do physical laws make things happened? *NetFuture, 155.* http://www.netfuture.org/2004/Mar1604_155.html

Tenner, E. (1997). *Why things bite back.* New York: Vintage Books.

Thoreau, H. D. (1999). *Material faith: Henry David Thoreau on science* (L. D. Walls, Ed.). Boston: Houghton Mifflin Company.

Troll, W. (1957). *Praktische Einführung in die Pflanzenmorphologie* [Practical introduction to plant morphology]. Jena: Gustav Fischer Verlag.

Wagenschein, M. (2009). Teaching to understand: on the concept of the exemplary in teaching. http://www.natureinstitute.org/txt/mw/exemplary_full.htm

West-Eberhard, M. J. (2003). *Developmental plasticity and evolution.* New York: Oxford University Press.

Whitehead, A. N. (1967). *Science and the modern world.* New York: Free Press.

Wilbur, H. M. (1976). Life history evolution in seven milkweeds of the genus Asclepias. *The Journal of Ecology, 64,* 223–240.

Wilber, K. (2000). *Sex, ecology, spirituality* (collected works volume 6). Boston: Shambala Publications.

Willson, M. F., & Rathcke, B. J. (1974). Adaptive design of the floral display in Asclepias syriaca L. *American Midland Naturalist, 92,* 47–57.

Wolf, G. (1996) A history of vitamin A and retinoids. *FASEB Journal, 10,*1102–7.

Woodson, R. E. (1954). The North American species of *Asclepias* L. *Annals of the Missouri Botanical Garden, 41,* 1–211.

Wright, J. P., & Jones, C. G. (2006). The concept of organisms as ecosystem engineers ten years on: progress, limitations, and challenges. *BioScience, 56,* 203–209.

Wyatt, R., & Broyles, S. B. (1994). Ecology and evolution of reproduction in milkweeds. *Annual Review of Ecology and Systematics, 25,* 423–441.

Wyatt, R., Broyles, S. B., & Derda, G. S. (1992). Environmental influences on nectar production in milkweeds (*Asclepias syriaca* and *A. exaltata*). *American Journal of Botany, 79,* 636–642.

Wylie, R. (1951). Principles of foliar organization shown by sun-shade leaves from ten species of deciduous dicotyledonous trees. *American Journal of Botany, 38,* 355–361.

Yang, R., & Shadlen, M. N. (2007). Probabilistic reasoning by neurons. *Nature, 447,* 1075–1080.

Ye, X., Al-Babili, S., Klöti, A., Zhang, J., Lucca, P., Beyer, P. et al. (2000). Engineering the provitamin A (beta-carotene) biosynthetic pathway into (carotenoid-free) rice endosperm. *Science, 287,* 303–5.

Zajonc, A. (1999). Goethe and the phenomenological investigation of consciousness. In S. R. Hameroff, A. W. Kasniak, & D. J. Chalmers (Eds.), *Toward a Science of Consciousness III* (pp. 417–427). Cambridge, MA: The MIT Press.

Acknowledgments

While I focused on the plant as a teacher in this book, here I would like to thank those human teachers who stimulated my interest in plants and helped me on my pathway toward a self-reflective and living understanding of nature: Jochen Bockemühl, Ernst-Michael Kranich, Wolfgang Schad, Andreas Such-antke, and Frank Teichmann.

Soon after I completed this book I learned that Henri Bortoft had passed away. Henri's thinking was truly alive, and he helped many people, myself included, move — as he would say — upstream to catch the world in its becoming. Thank you, Henri.

A Ph.D. dissertation on "Living Thinking for a Culture of Transformation" went through substantial metamorphosis involving much shedding and new formations to finally become this book. I want to thank all the people who contributed to the development of the book by reading and commenting on chapters or the whole manuscript at some stage: Linda Bolluyt, Christina Holdrege, Henrike Holdrege, Michael Holdrege, Rick Medrick, Pramod Parajuli, Mark Riegner, Douglas Sloan, Steve Talbott, and Arthur Zajonc. I am grateful to Will Marsh for his helpful copy editing and to Mary Giddens for doing the layout of the book.

My work on this book was supported by The Nature Institute, which can carry out its non-profit efforts due to the generous support of individuals and foundations, including the Foundation for Rudolf Steiner Books, Kalliopeia Foundation, GLS Treuhand, Mahle Foundation, Rudolf Steiner Charitable Trust of RSF Social Finance, Rudolf Steiner-Fonds, Salvia Foundation, Software AG Foundation, Stiftung Evidenz, and Waldorf Educational Foundation. My thanks to all of these organizations.

I would like to thank the many participants in Nature Institute courses and workshops; without the work with other people and the collaborative exploration of plants and knowing, this book would never have developed.

Finally, special thanks go to Henrike, my wife and colleague. You put up with a longer-than-planned process and gave me encouragement in the times when I felt stuck and frustrated. And now you can rejoice that it has finally come to fruition.

Craig Holdrege

Index